Chemistry and Enzymology
of
Marine Algal
Polysaccharides

Himanthalia lorea and *Laminarias* at
Dancing Ledge, Dorset.

Chemistry and Enzymology
of
Marine Algal
Polysaccharides

ELIZABETH PERCIVAL

Department of Chemistry
Royal Holloway College (University of London)
Surrey, England

and

RICHARD H. McDOWELL

Alginate Industries Ltd
London, England

1967

ACADEMIC PRESS
LONDON AND NEW YORK

ACADEMIC PRESS INC. (LONDON) LTD
Berkeley Square House
Berkeley Square
London, W.1

U.S. Edition published by
ACADEMIC PRESS INC.
111 Fifth Avenue
New York, New York 10003

Library of Congress Catalog Card Number: 67–21943

Made and printed in Great Britain by
William Clowes and Sons, Limited, London and Beccles

Preface

The long history of the use of seaweeds for a variety of purposes has led to the gradual realization that their constituents are somewhat different in composition from those of land plants. They synthesize a wide variety of fascinating polysaccharides, some of which, such as the algal starches, closely resemble those of the higher plants, whereas the sulphated galactans and fucoidan are unique to the red and brown seaweeds, respectively. The general nature of the main polysaccharides of the common marine algae has been known for many years, and can be found from numerous works of reference, but our knowledge of the fine structure of these compounds is largely the result of investigations during the last decade. During that time also many more species of algae have been examined, and in addition various minor carbohydrates of well-known genera have been studied. Although we are very far from understanding the complete metabolic processes involved in the life cycle of the algae, important studies in the last few years provide a useful basis for further work.

The emphasis in this book is on the structure and the general relationships of the polysaccharides as they occur in the algae, rather than on the commercial products from seaweeds, although some space is devoted to the properties and uses of alginates, agar and carrageenan, and comprehensive references to all aspects are included.

Attention is also given to the studies which have been made of the action of enzymes from various sources on some of the algal polysaccharides, and the products which are derived from them. Recent biosynthetic studies involving nucleotides are also described.

While some of the polysaccharides have received a great deal of attention from research workers, others have been largely neglected. This is reflected in the emphasis given in the different chapters.

The book is intended for chemists and biologists and in particular for those engaged in any aspect of phycological research, and should be of value in explaining some of the changes which take place in the living algae during development. Industries which make use of plant gums and mucilages should also find material of value to their particular problem.

A knowledge of general organic and physical chemistry is presumed but Chapter 2 is devoted to a detailed explanation of all the recent techniques that have been applied to the elucidation of the structure of these polysaccharides thus avoiding repetition of the details when discussing particular polymers.

We should like to thank Professor D. J. Manners and Dr. D. A. Rees for helpful criticisms of the chapters on laminaran and on the galactans, respectively, and to record our gratitude to Professors C. Araki, R. G. S. Bidwell, W. Z. Hassid and T. Miwa and to Drs. A. Haug, J. R. Turvey and W. Yaphe for permission to include some of their unpublished results. We also thank Dr. J. H. Price of the British Museum (Natural History), Mr. E. Booth of the Institute of Seaweed Research, and Alginate Industries Ltd. for photographs used as illustrations.

While correcting proofs, and in order to bring the work completely up to date, the opportunity has been taken of referring to some papers published over the last few months.

April, 1967   *Elizabeth Percival*
                                                       Richard McDowell

Contents

CHAPTER 4

OTHER NEUTRAL POLYSACCHARIDES, FOOD RESERVE AND STRUCTURAL

CHAPTER 5

1*

CHAPTER 6

Sulphated Polysaccharides Containing Neutral Sugars: 1.

Galactans of the Rhodophyceae 127

CHAPTER 7

SULPHATED POLYSACCHARIDES CONTAINING NEUTRAL SUGARS: 2 . . 157

CHAPTER 8

POLYSACCHARIDES CONTAINING URONIC ACID AND ESTER SULPHATE . 176

CHAPTER 9

CHAPTER 1

Polysaccharides in Living Marine Algae

I. INTRODUCTION

The quantity of polysaccharides synthesized by plants in the oceans probably exceeds that found on land and in fresh water, and, unlike fresh water in which flowering plants are common, the sea has a vegetation consisting almost entirely of algae. They have a great diversity of form, ranging from unicellular organisms to the giant seaweed *Macrocystis pyrifera* which may attain up to sixty metres in length, and *Lessonia flavicans* which grows to the size of small trees. Being essentially photosynthetic organisms, they grow only in those parts of the sea where light is available. The free-floating types, most of them unicellular, can be found in the upper layers of all the seas, but most of the larger species are attached to rocks and are, therefore, found only in the continental shelf areas, more particularly in those parts where the sea bottom is sufficiently stable to provide a firm anchorage. As this type of habitat is very small compared with the total area of the oceans, by far the greater weight of marine algae in existence is made up of the unicellular forms, but on the other hand the larger plants near the shores are far more conspicuous, and have been more widely studied.

The economic importance of the unicellular species is mainly indirect as they form the first stage in the food cycle of life in the sea, but those on and near the shores of many countries have been used for centuries either as human food, as food for animals, as manure, or as a source of chemicals.

As with land plants, the predominant algal species in any area depends on the environment, and there is a very noticeable zonation on those shores where there is a considerable tidal range. The areas which are frequently exposed are occupied mainly by green seaweeds, and members of the order Fucales of the brown seaweeds, usually with further zonation by species. Other species of brown algae generally predominate from immediately below low tide level to a depth of ten to twenty metres, although in some areas red seaweeds are more important. With increasing depth, fewer plants are present; mainly red species. The greatest depth at which algae have been found is about two hundred metres (Smith, 1955a). An important factor in the zonation of plants with depth is, no doubt, the change in spectral distribution of light penetrating the water, those algae with pigments suitable for making use of the light being the most abundant (Levring, 1947). It will, however, be realized that some species from each of the classes are to be found at all tide levels and that other factors besides depth of immersion are involved in the ecology of the algae (Biebl, 1962).

II. CLASSIFICATION AND THE MAJOR POLYSACCHARIDES

Of the thousands of species of marine algae which have been described—and no doubt many more have yet to be discovered—only a small proportion have been examined chemically. Any generalizations about the nature of the polysaccharides of different classes of algae can therefore only be tentative.

The botanical classification of the seaweeds is based primarily on their morphology, particularly in respect of the reproductive systems, but the division into classes has been aided by the nature of the pigments present. All marine algae contain chlorophyll *a*, but the colour of the chlorophyll in some classes is masked by strongly coloured pigments. For example, nearly all species of the Rhodophyceae are coloured bright red by biliproteins, and the Phaeophyceae derive their brown colour from the xanthophyll, fucoxanthin. Some other pigments have been used to differentiate algae in other classes (Strain, 1951).

Although classification is being modified as more species are examined, the classes, Phaeophyceae (brown seaweeds), Rhodophyceae (red seaweeds) and Chlorophyceae (green seaweeds) are well established, and there is general agreement about many of the subdivisions within them (Parke and Dixon, 1964; Silva, 1962). Algae from these classes are the most easily obtainable and most directly important economically, and include the greater number of species which have been examined chemically. Each class has its particular types of polysaccharide as well as its characteristic pigments, although in some cases the fine structure of the former varies from species to species.

Many of the polysaccharides of the algae, although different in detail, have their counterparts in land plants. For example small proportions of cellulose are found in some species of each of the three major classes mentioned above. On the other hand, a general characteristic of marine algae appears to be the presence of at least one polysaccharide linked with sulphate ester groups. These substances are not found in land plants, although complex sulphated polysaccharides are present in animal tissues.

Although from time to time the presence of chitin has been reported in algae, largely on the basis of staining reactions, chemical investigations to show its presence have given negative results (Young, 1966) except in the diatoms, *Thalassiosira fluviatilis* and *Cyclatella cryptica* (Falk *et al.*, 1966).

A. POLYSACCHARIDES OF THE PHAEOPHYCEAE

Most of the brown seaweed species which have been examined contain the β-1,3-linked glucan, laminaran (Quillet, 1958), and where it is absent neither starch nor any other glucan has been found. The polyuronide, alginic acid, is present in all the species so far examined, and apart from the formation of a similar compound by bacteria grown under artificial conditions, no other source of this polysaccharide has been found. Variations in the proportions of the two types of residue present, namely mannuronic and guluronic acid, are found in samples of alginic acid from different species (Fischer and Dörfel, 1955), but these seem to bear no relation to the recognized subdivisions of the Phaeophyceae.

All the species of brown algae examined contain a sulphated polysaccharide typified by the presence of the sugar fucose. In most cases some other sugars and perhaps polyuronides and protein are combined in the molecule (see p. 176) but too few samples from different species have been examined to give any indication of the relation of composition to classification.

B. POLYSACCHARIDES OF THE RHODOPHYCEAE

Floridean starch can be observed by microscopic examination of species of Rhodophyceae stained with iodine, and in those cases where it has been examined chemically, it has been found to be an α-1,4-linked branched glucan similar to the amylopectin of land plants.

The main polysaccharides of most red seaweeds which have been examined are galactans. Units in them are D- and L-galactose (some carrying half ester sulphate), 3,6-anhydro-D- and L-galactose, and 6-O-methyl-D-galactose. A chain of alternating 1,3- and 1,4-glycosidic links seems to be general, but this allows a number of different types of polymer to be built up. Although it seems probable that each species contains more than one type, there is fairly good agreement between the botanical classification and the nature of the galactan present. A classification based solely on the properties of the galactan has been suggested by Stoloff and Silva (1957), which, although it largely agrees with the botanical system, suggests minor alterations. As the chemical examination of some of the compounds has not yet been completed, it would seem premature to make modifications.

The polysaccharide which is most easily extracted from *Rhodymenia palmata* is a xylan; whether a galactan is also present has not yet been determined. It should also be noted that a xylan (see p. 90), and a mannan (see p. 93) have been isolated from *Porphyra umbilicalis* in which the main polysaccharide is a galactan.

C. Polysaccharides of the Chlorophyceae

The Chlorophyceae (green algae) include many fresh-water species, but most of the structural studies on the polysaccharides have been made on those from the sea. It is probable that they all contain starch similar in composition to that from land plants comprising both amylose and amylopectin (see p. 79), but the starch granules in the algae are less well organized than those in the higher plants.

All the species that have been examined contain a high proportion of complex water-soluble sulphated polysaccharides. Those from *Ulva*, *Enteromorpha* and *Acrosiphonia* have L-rhamnose, D-xylose and D-glucuronic acid as their main constituents, while *Cladophora*, *Chaetomorpha*, *Caulerpa* and *Codium* have polysaccharides all based on D-glactose, L-arabinose and D-xylose, but in rather different proportions. These two groups of polysaccharides are not entirely in line with the botanical classification, as *Acrosiphonia* (*Spongomorpha*) is at present considered to be closely related to *Chaetomorpha* and *Cladophora*. More work on a greater variety of species is clearly needed before the information can be of great value in assisting with classification.

D. Polysaccharides of Other Classes of Algae

Polysaccharides have been isolated from marine algae of some other classes, such as the Cyanophyceae (blue-green) and the Bacillariophyceae (diatoms), but the number investigated has been too small to draw any conclusions regarding classification.

III. LOW MOLECULAR WEIGHT CARBOHYDRATES IN MARINE ALGAE

Before considering the metabolic processes of the algae and the variations in their composition, some attention must be paid to their low molecular weight carbohydrates as well as to their polysaccharides.

Each of the main groups of marine algae contain characteristic low molecular weight carbohydrates, some of which appear to be chemically related to the polysaccharides of that group. The distribution of these substances has been reviewed by Lindberg (1955a) and by Meeuse (1962). While some of these low molecular weight materials account for a considerable proportion of the dry weight of the alga, others are present only as trace quantities and their identification has been possible only by paper chromatography.

A. In the Phaeophyceae

D-Mannitol appears to be universally present in the brown algae, in some species and at some seasons in large quantities (about 25% of the dry weight of some *Laminaria* species in the autumn) (Black, 1950a). The seven carbon atom sugar alcohol, D-volemitol (1) was found in addition in *Pelvetia canaliculata* (Lindberg and Paju, 1954). Furthermore, 1-*O*-D-mannitol-β-D-glucopyranoside and 1,6-*O*-D-mannitol di-(β-D-glucopyranoside) (2) were found in all the brown algae investigated by Lindberg and co-workers, while β-glucosides of D-volemitol were found in *P. canaliculata*. A c-methylinositol, laminitol, was found in *Laminaria hyperborea* and related species, and trace quantities of sucrose, galactose and mannose have been reported in *Cladostephus* species (Fanshawe and Percival, 1958).

B. In the Rhodophyceae

A galactoside, given the name floridoside when first isolated (Colin and Guéguen, 1930), was later shown to be 2-*O*-glycerol α-D-galactopyranoside (3) (Putman and Hassid, 1954). It is present in many of the Rhodophyceae, in high proportions in some (Majak *et al.*, 1966). In other species 2-D-glyceric acid *a*-D-mannopyranoside (4) is more important (Colin and Augier, 1939). In addition, 3-*O*-floridoside α-D-mannopyranoside (5) has been isolated from *Furcellaria fastigiata* (Lindberg, 1955b) and 1-*O*-glycerol α-D-galactopyranoside (*iso*-floridoside) (6) (Lindberg, 1955c) from *Porphyra umbilicalis* and other red weeds (Majak *et al.*, 1966).

Various sugar alcohols and inositols have also been found in some species of red algae.

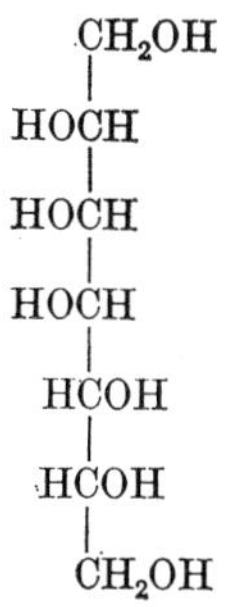

(1) 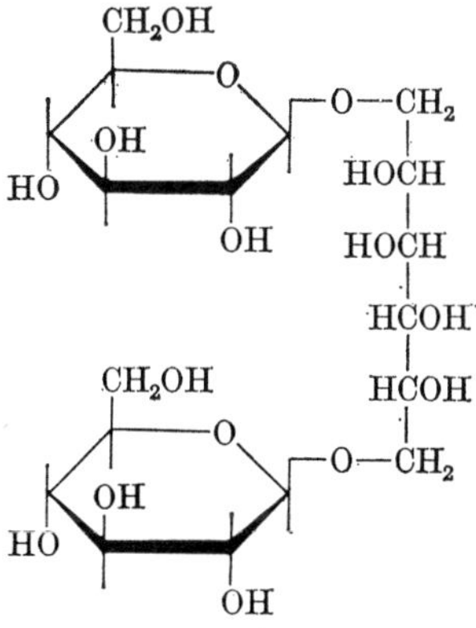   (2) 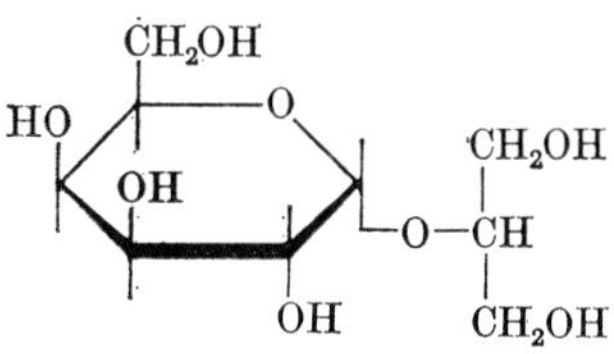   (3)

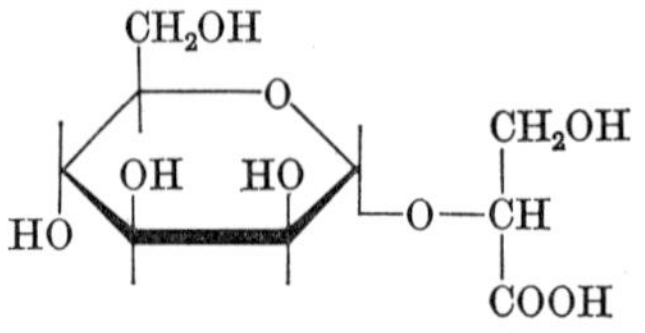

(4)

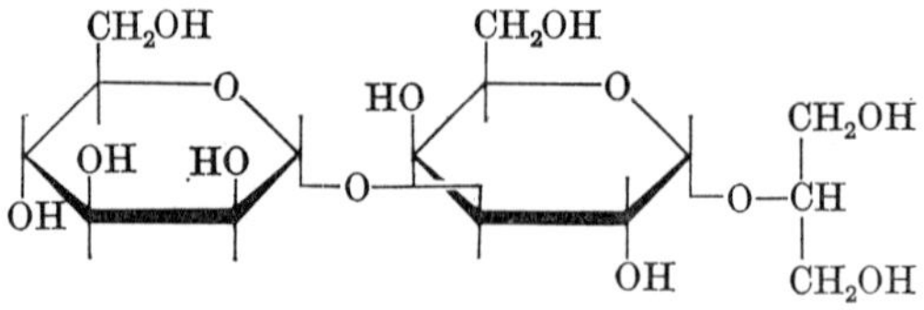

(5)

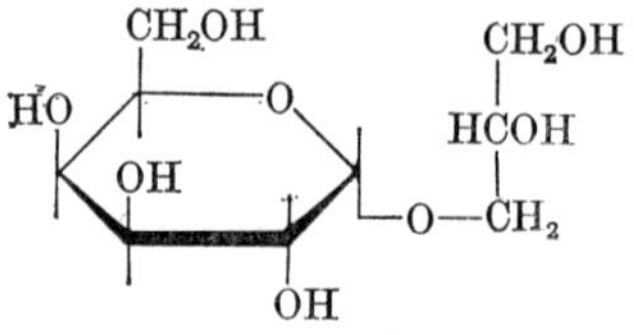

(6)

C. In the Chlorophyceae

This class of algae has not been investigated to the same extent as the Rhodophyceae and the Phaeophyceae, but sucrose, glucose and fructose have been found in many of the genera (Craigie et al., 1966).

D. In Other Classes of Algae

Mannitol was found to be the major product of photosynthesis in *Olisthodiscus* (Xanthophyceae) (Bidwell, 1957), and glucose together with laminitol and other inositols and some oligosaccharides were isolated from the diatom, *Phaeodactylum tricornutum* (Bacillariophyceae) (Ford and Percival, 1965a).

IV. VARIATIONS IN THE CARBOHYDRATE CONTENT OF ALGAE

Extensive studies have been made on the factors affecting the composition of the brown algae. For many species the composition has been shown to depend on the season of the year, the habitat, depth of immersion, and state of development. Most work has been done on species of commercial importance, for example, the Laminariales (Black, 1950a) and the Fucales (Black, 1948, 1949; Macpherson and Young, 1952). The figures are generally quoted on the basis of the proportions of the different constituents in the total dry solids, which is the most informative method from the point of view of the economic use of the seaweed, but other factors must be considered in studying the changes during synthesis and utilization of the carbohydrates by the plants. For example, with the common British Laminariales, *L. hyperborea*, *L. digitata* and *L. saccharina*, the proportions of laminaran and mannitol in the dry solids increase steeply over the period April to September, and at the same time both the percentage of dry solids in the wet weight and the weights of the plants are increasing. On the other hand, the proportion of alginic acid and of cellulose in the dry matter decreases in that period, and increases from October to April while the proportion of total dry solids is decreasing. During this time there is little growth and most of the plants decrease in weight as a result of breakage by storms. Variations are much greater in the fronds, which have high growth rates, than in the stipes where growth is slower.

In most of the analyses which have been reported, whole fronds or complete plants have been used, and they have therefore been on a mixture of tissues of different ages. It has been shown (Black, 1954a) that in a mature frond of *L. saccharina*, where the part near the tip was about seven months old, there was marked variation in composition along the length. Near the stipe (i.e. the actively growing region) there was, on a fresh weight basis, about 3% mannitol and little or no alginic acid or laminaran. About a third of the way along the frond mannitol was at a maximum of about 6% with laminaran 2% and alginic acid 2·5%; while two-thirds of the way up the frond the mannitol content was only 2% with laminaran at 6% and alginic acid 4%. Variations in the composition of the whole fronds during the period of rapid growth can therefore be due largely to changes in proportions of old and new tissues. A similar dependence of the composition on the parts of the plant analysed has also been found in *Fucus vesiculosus* (Moss, 1948) and *Himanthalia elongata* (Jones, 1956).

A further point is the variation in composition among a number of plants collected in the same area at the same time (Black *et al.*, 1959). Determination of the mannitol and ash contents of the stipes and of old and new fronds of *L. hyperborea* showed such variations from plant to plant that it was concluded that seasonal variations found in the stipes are not statistically significant. Although seasonal variations of these constituents in the fronds are large enough to be significant, the results as a whole are an indication that small seasonal variations which have been reported for some constituents of algae need further investigation before any metabolic significance is assumed.

A. Factors Influencing the Variations in Composition

The amounts of different constituents in the seaweed can be regarded as the result of a number of enzymic reactions and diffusion processes, the rates of which are influenced in different ways by the many variables affecting the plant.

Variations in composition of the algae are conveniently studied by sampling beds of seaweed at different times of the year and at different depths and localities, and the results are of the greatest value to the user of seaweed in terms of these variables. On the other hand, the primary factors influencing the rate of growth, and hence the composition of the plants, are the temperature of the water, the amount of light available for photosynthesis and the concentration of the nutrients in the water. These factors will vary with the locality and with the season of the year, and can be expected to have somewhat different values from one year to another. Furthermore, they are interrelated as the amount of sunshine will affect both the light available for photosynthesis and the temperature of the water, and the concentrations of nutrients are highly dependent on movements of different layers of water resulting from changes in temperature.

Studies on the effect of depth of immersion on the composition of *Laminaria* species (Black, 1950b) are in agreement with previous suggestions that the maximum photosynthesis takes place at a depth of six to ten metres and not at the surface, at least in periods of bright sunshine (Levring, 1947).

The very widespread increase in the rate of growth in algae in the spring is due partly to increases of temperature and amount of light, but in the case of diatoms it is well established that the main cause of higher growth rates in the spring than in the summer is a good supply of nutrients, particularly phosphate and nitrate, brought about by upwelling of deeper water with changes of temperature during the winter (Harvey, 1955). Some correlation between the rate of growth of the *Laminarias* and the amount of phosphate and nitrate in the sea has been found (Black and Dewar, 1949), but the effect on the production of the various constituents of the plants is dependent on a number of processes which are imperfectly understood.

B. Seasonal Variations in the Laminariaceae

The seasonal variations reported for the Laminariaceae appear to be consistent with the following:

(*a*) Mannitol is the first product of photosynthesis to accumulate in appreciable quantities and is the main carbohydrate in tissues which are increasing by active cell division.

(*b*) During further photosynthesis in tissues which are growing largely by cell enlargement, there is an increase in the proportion of the dry solids made up partly of more mannitol but also of salts of alginic acid and of laminaran. Cellulose and protein are also being synthesized. Formation of mannitol and laminaran continues after the other constituents have built up to a constant level in each unit of tissue, thus increasing the dry solids content and reducing the proportion of alginate, cellulose and protein on the dry weight basis. Thus in late summer mannitol and laminaran are at high levels and alginic acid, cellulose and protein are at a minimum on this basis.

Laminaria hyperborea stipe and frond.

Laminaria hyperborea at low tide.

(c) The laminaran may be formed from mannitol, so that during active growth the mannitol can be formed faster than it is converted into laminaran and both substances increase in amount. But when growth slows down or stops due to lack of nutrients, shortage of light or low temperatures, laminaran increases with loss of mannitol. (For evidence on interconversion of mannitol and laminaran, see p. 19.) Thus, in late summer there may be a temporary reduction in mannitol content due to depletion of phosphate in the water, while the laminaran content does not drop until later.

(d) During spore formation and periods when respiration is greater than photosynthesis, both laminaran and mannitol are used up. As there is little change in the amounts of other constituents, the proportions of alginate, cellulose and protein, calculated on a dry basis, increase. Results are not sufficiently precise to determine whether the reduction in the proportion of total solids is accounted for entirely by the consumption of laminaran and mannitol or whether there is a change in the amount of water held by the tissues.

C. Variations in the Fucaceae

Although the Fucaceae contain the same constituents as the Laminariaceae, it is not surprising that, with their rather different environments (involving much more exposure to the air than is the case with the latter) the proportions and seasonal variations in their carbohydrate constituents are rather different.

Fucoidan, which is a minor constituent in the Laminariaceae, is present in larger amounts than mannitol and laminaran in some of the Fucaceae. The quantity present is largely dependent on the degree of exposure to the air; *Fucus spiralis* and *Pelvetia canaliculata* which grow in the highest zone on the shores contain the largest amounts (10 to 13% fucose, corresponding to about 18 to 24% fucoidan on a dry basis). On the other hand, *Fucus serratus* which grows round about the low water mark has much less (not more than 7% fucose, that is about 13% fucoidan on a dry basis) (Black, 1954b).

Seasonal variations in the composition of the Fucaceae are not so marked as in the Laminariaceae, probably because the rate of growth is much slower.

D. Variations in Other Algae

There has been very little systematic work on variations in composition of other classes of marine algae. A number of different samples of red algae were examined as a source of carrageenan (Black *et al.*, 1965) and some indications of seasonal variation were found in both amount and composition, but suitable samples were insufficient to make any generalizations possible.

There appears to be some correlation between the 3,6-anhydrogalactose content of the porphyran obtained from different samples of *Porphyra perforata* and the degree of exposure to wave action, but samples of *Porphyra umbilicalis* collected at different times of the year showed no significant variations in composition (Rees and Conway, 1962).

V. LOCATION OF POLYSACCHARIDES IN THE ALGAE

The location of substances in plant tissues can give some indication of their function, for example, carbohydrates found inside the cells can be considered as playing an active part in metabolism. In the higher plants, reserve materials are commonly located within the cells of specialized tissues removed from the site of photosynthesis. On the other hand, those polysaccharides which are found in the cell walls can be described as structural.

In addition there are many plant polysaccharides which are found outside the cells, and not forming part of the cell wall. In the higher plants they are classed mainly as pectins, gums and mucilages; their functions are not very well understood but in some cases they may add to the mechanical strength of the tissues.

The algae are generally considered to be lower in the evolutionary scale than the land plants and they show much less differentiation into tissues with specialized functions. As they live all or part of their time completely surrounded by water, a much simpler structure is adequate for their growth and reproduction than that of land plants, which are adapted to have part in the soil and part in the air. Although there may be some connections between the cells of the multicellular algae, they do not have the vascular tissues typical of land plants, and the only evidence for translocation of nutrients is in *Macrocystis pyrifera* (Clendinning, 1964). Most species have specialized reproductive cells, and the actively growing region may have a distinctive structure; some modification of the plant is also common in the holdfast by which the plant is held to a rock, and there may be a distinctive stipe between the holdfast and the main part of the plant. At the same time, the outer part of the tissues may have a rather different structure from the interior. Where these differences are marked, for example, stipe and frond of *Laminaria* species, separate analyses of the carbohydrates of the different tissues have revealed variations in the quantities more than in the nature of the constituents (Black, 1954a; Moss, 1948; Jones, 1956). There appears to be no differentiation into photosynthetic and storage tissues, and the use of whole algae, which have in most cases been the raw material for the study of their constituents, can therefore be justified.

A. IDENTIFICATION BY DIFFERENTIAL EXTRACTION

A commonly used method of studying plant materials is to make successive extractions with conditions so arranged that different groups of compounds are dissolved at the various stages (see p. 26). The ease of extraction will depend partly on the location of the materials in the plant and partly on their chemical nature. Cell contents can often be separated from other materials by mechanical means and more completely by cold water, but some seaweeds also exude a water-soluble mucilage which is thought to originate from the intercellular spaces rather than from the interior of the cells. The basic cell wall is generally considered to be the material remaining undissolved after extraction with hot dilute solutions of acid and alkali or oxidizing agents; in some cases this process can be continued until the residue consists wholly or almost entirely of a single substance. Cellulose is isolated in this way from most tissues of higher plants and from some algae (p. 84). While some siphonaceous green algae yield a xylan (Iriki et al., 1960; Mackie and Percival, 1959), others again have a mannan as this residual material (Iriki and Miwa, 1960; Love and Percival, 1964a) (see p. 94). The methods used by the different investigators vary in detail, but an essential point is that the extracted material, where examined, differed in composition from the residue.

Although diatoms are well known for the siliceous skeletons which remain after organic matter has decayed, polysaccharides form an important part of the cell wall in the living organisms. From freeze-dried cells of the diatom, *Phaeodactylum tricornutum*, a residue was obtained after treatment with hot water, chlorite, and cold 4% sodium hydroxide; this contained 18% of carbohydrate which proved to be a sulphated glucuronosylmannan (Ford and Percival, 1965b) (see p. 186). In this case again, the earlier extracts contained different carbohydrates.

On the other hand, some of the carbohydrates which are removed from the tissues by this exhaustive extraction probably form part of the cell wall, and may in some cases constitute the major part of it (Frei and Preston, 1961).

Differential extraction does not always lead to the separation of a series of polysaccharides from a final residue of different composition. Successive extractions of dried *Enteromorpha compressa* with hot water, dilute chlorous acid, 4% sodium hydroxide and 18% sodium hydroxide gave extracts of similar composition; all of them and the small amount of residual weed contained polysaccharides which gave on hydrolysis rhamnose, xylose, glucuronic acid and glucose. They differed in that the first three extracts contained a major proportion of rhamnose while in the last extract and the weed residue glucose was present in the largest amount (McKinnell and Percival, 1962).

Similarly, in *Rhodymenia palmata* a xylan is the major constituent in the aqueous extract while the residue after successive extractions contains xylose and glucose. Whether the same xylan is present in both parts or whether the xylose is linked with glucose in the residue has not yet been established.

Chemical examination of the extracts and the residues at different stages of the treatment is not sufficient to decide their original location, but is of great value when combined with various physical examinations of the extracted residues (Cronshaw et al., 1958).

B. IDENTIFICATION BY OPTICAL MICROSCOPE

By far the greater part of the study of plant tissues has been made with the optical microscope, generally using stained sections of unmodified material, but also making use of polarized light. Tissues from which some constituents have been extracted have also been studied, for example, sections of Laminaria before and after the alginate had been dissolved (Thiele and Andersen, 1955).

The unique staining reaction of starch by iodine together with its behaviour in polarized light have enabled the site of the starch in green and red seaweeds to be determined by microscopic examination. It is present in close association with the synthetic pigments, but while in the green seaweeds it is mainly found in the form of granules in the chloroplasts (Smith, 1955b), in the red seaweeds the somewhat different floridean starch (see p. 74) is distributed through the cytoplasm in the form of layered granules which are thought to have formed on the outer surface of the chromatophore (Meeuse et al., 1960).

No specific stain has been found for laminaran, so that although it is probably located within the cells of the Phaeophyceae in the same way as starch in other algae, its presence there has not been demonstrated by microscopic methods.

Both the cell walls and the mucilaginous region between the cells are stained by a number of reagents, toluidine blue O and periodic acid-Schiffs (PAS) reagents (McCully, 1965) being particularly useful. Although there are some differences of tint with toluidine blue with polysaccharides containing sulphate groups and those with uronic acid groups, it is difficult to draw clearcut conclusions when both types of polysaccharide are present. A positive PAS reaction is a good indication of polysaccharides with adjacent free hydroxyl groups in the residues, so that cellulose and alginic acid are coloured by it.

Insufficient is known of the structure of the sulphated polysaccharides of the brown seaweeds to be certain whether they would stain or not. The use of non-specific stains, such as haemotoxylin, can also be of value in the microscopic examination of sections of algae (Baardseth, 1966).

C. Examination of Cell Walls by X-ray and Electron Microscope

Studies have been made on the cell walls of a number of species of algae, particularly of the Chlorophyceae, using X-ray and electron microscope techniques, generally after preliminary chemical treatment of the material (Frei and Preston, 1961). In many cases distinct microfibrils are demonstrated by the electron microscope, but this is not universal, nor do the microfibrils from all the algae have the same chemical composition. In most of the Rhodophyceae and Phaeophyceae which have been examined, the X-ray diagram of cellulose I (the native cellulose of land plants) is shown after the microfibrillar material has been freed from encrusting materials of the cell wall, but in *Porphyra umbilicalis*, xylan and mannan are present in the cell wall (Peat *et al.*, 1961), while cellulose is absent. X-ray and electron microscope evidence shows that the microfibrils in this case are composed of a xylan while the mannan is present as an outer layer or cuticle of the frond in a less highly orientated form (Frei and Preston, 1964b).

Many of the Chlorophyceae, for example, *Valonia* which has been extensively investigated, have cell walls with an abundance of highly crystalline cellulose I in the form of relatively straight and thick microfibrils, while, as mentioned above, the crystalline material in other species may be a xylan or a mannan. The chemical evidence for this has been confirmed by X-ray analysis (Frei and Preston, 1964a).

The X-ray evidence is of the greatest value when the plant material can be treated to yield a single polysaccharide in a crystalline form; combined with electron microscope photographs, the structural function of the polysaccharides so studied has been amply demonstrated (Preston, 1965). On the other hand, with tissues which contain a number of different materials, interpretation of the X-ray diagrams is difficult and chemical work is likely to be more reliable.

D. Encrusting Cell Wall Materials and the Intercellular Matrix

In the multicellular algae, the proportion of the weight which is contributed by mucilaginous matter is generally very much greater than is the case with most land plants. In the red and brown seaweeds, such materials often account for more than 30% of the dry weight of the plant, and more by volume in the highly swollen state in which they are present in the living plant.

Whether some of these constituents are considered as part of the cell wall or to be in a region between the cell walls is partly a matter of definition. The cell walls certainly consist of more than the microfibrils; there is some less well organized material closely attached to them, and it is probable that some protein plays an essential part in the growth of the cell walls (Preston, 1965). It seems not unlikely that polysaccharides with uronic acid or sulphate groups, universally present in algae, are associated with this protein (see p. 176).

As more detailed examination is made, the great complexity of cell wall structures is becoming increasingly apparent. For example, it has recently been claimed that the cell walls of the green seaweed, *Prasiola japonica*, are made up of at least four layers of different chemical composition (Takeda *et al.*, 1967).

Some algal polysaccharides are so easily extracted by treatment of the tissues with cold water that it is hardly likely that they form part of the cell wall. This, together with evidence from microscopic examination of sections, makes it reasonably certain that the sulphated polysaccharides from red and brown seaweeds are located between the cells to form a viscous matrix. In some other species the water-soluble materials may come from either the intercellular region or from inside the cells. The water-soluble material from the green seaweed, *Codium fragile* (which is unicellular although forming plants up to at least forty centimetres long) is a sulphated polysaccharide containing galactose and arabinose residues (Love and Percival, 1964b). The rather similar extracts from *Cladophora rupestris* and *Chaetomorpha capillaris* might, therefore, also originate from within the cells.

VI. METABOLIC STUDIES AND ENZYME SYSTEMS IN ALGAE

While much of our knowledge of photosynthesis and carbon dioxide metabolism has come from work on species of *Chlorella* and other fresh-water unicellular algae, it is apparent from their different composition that the red and brown seaweeds must have rather different metabolic systems, but very little is known about them. The experimental difficulties of working with multicellular marine algae compared with unicellular fresh-water algae are very considerable, and only a few experiments on the growing plants have been made. Studies on the enzyme systems of algae have also proved difficult as the extracts so far obtained have been far less active than those from land plants, microorganisms and animal tissues. Direct metabolic studies and work with enzymes extracted from marine algae are discussed in this section; the action of enzymes from other sources on individual seaweed polysaccharides is dealt with in their particular sections.

A. PHOTOSYNTHESIS AND THE FORMATION OF LOW MOLECULAR WEIGHT CARBOHYDRATES

Although the algae contain a variety of pigments, it is considered that chlorophyll *a* is solely responsible for the photosynthetic process, and the importance of other pigments lies in their power to absorb light of other wavelengths, the energy then being transferred to the chlorophyll *a* (Brody and Brody, 1962). The light reaction of photosynthesis in the algae is therefore the same as that in higher plants. Similarly there is no reason to suppose that the first stage in the assimilation of carbon dioxide is any different from that of other plants; it is in the later stages of synthesis that the differences are found.

In a study of the uptake of radioactive carbon by the red alga, *Iridaea* (*Iridophycus*) *flaccidum*, during photosynthesis using $^{14}CO_2$ (Bean and Hassid, 1955) the first radioactive compound found with short periods of illumination was 3-phosphoglyceric acid, as is the case with land plants. Radioactive floridoside was first observed in the experiment after thirty seconds illumination, and after two hours was the main radioactive product with very slight radioactivity in the alcohol insoluble fraction. It formed a slightly higher proportion of the alcohol soluble radioactive products after ten hours illumination, but by that time the proportion of radioactivity in the alcohol insoluble fraction had risen to about 20%.

Uridine diphosphate (UDP)-glucose appears in small amounts early in the photosynthetic process, followed shortly after by UDP-galactose, and it is concluded that the glucose derivative is an intermediate in the formation of the latter, which in its turn is utilized in the synthesis of floridoside, and of the galactans which form a large part of the plant. In contrast with land plants and green algae, no sucrose was found among the radioactive products.

When the photosynthetic products obtained from a series of marine algae from the classes Phaeophyceae, Rhodophyceae and Chlorophyceae during illumination in the presence of $^{14}CO_2$ were examined (Bidwell, 1958) no radioactive phosphoglyceric acid was found, but with the long periods of illumination used (four to six hours) it would be expected to be only a small proportion of the radioactive products. The materials present in the largest quantities varied with the class of seaweed, being mannitol in the Phaeophyceae, floridoside in the Rhodophyceae, and sucrose in the Chlorophyceae. *Polysiphonia lanosa* (Rhodophyceae) appeared to be exceptional in giving an unidentified carbohydrate which yielded mannose and glycerol on hydrolysis.

In a further study of the uptake of radioactive carbon dioxide by twelve algal species from six orders of the Rhodophyceae (Majak *et al.*, 1966), floridoside was again found to be the chief alcohol soluble radioactive carbohydrate. It was, however, absent or only present in small amounts in the

Ceramiales. *Batrachospermum* (Nemalionales), which contained trehalose ^{14}C, also lacked floridoside.

The photosynthetic products of *F. vesiculosus*, obtained under different conditions, were examined in more detail by Bidwell *et al.* (1958). Using different concentrations of $^{14}CO_2$ in the atmosphere and with varying amounts of nutrient salts in the medium, there was little difference in the proportions of the radioactive products. After thirty hours illumination, material insoluble in alcohol contained about a third of the activity, but mannitol contributed about 80% of the activity of the alcohol-soluble products. The findings on the alcohol-insoluble material are discussed below.

B. ENZYMIC SYNTHESIS OF POLYSACCHARIDES

Very little is known of the processes by which the major polysaccharide constituents of the algae are built up, nor indeed is there very much information about which low molecular weight substances are utilized in the syntheses.

Considerable evidence has accumulated in recent years that the glycosyl esters of nucleotides act as donors in polysaccharide synthesis (Neufeld and Hassid, 1963; Ginsberg, 1964).

1. *Nucleotides in the Phaeophyceae*

A number of nucleotides, among them guanosine diphosphate (GDP)-mannuronic acid and GDP-guluronic acid, have been isolated from the brown seaweed, *Fucus gardneri* (Lin and Hassid, 1964, 1966a).

The dried seaweed was extracted with 50% alcohol and after concentration of the extract, the nucleotides were adsorbed from it on charcoal–Celite. They were eluted from this with 50% ethanol containing 0·1% ammonia, adsorbed on a Dowex 1 × 8 column and fractions obtained by gradient elution with an ammonium formate–formic acid system. Fractions containing nucleotides were identified by UV-absorption at 260 mμ and 275 mμ. Sixteen major fractions were obtained in this way, one shown by the carbazole reaction to contain uronic acid. The UV-spectrum and chromatography of the hydrochloric acid hydrolysate showed the presence of guanosine in the uronic acid nucleotide fraction.

The products of mild hydrolysis of the nucleotides were used to investigate the uronic acids which were first separated from other materials by paper chromatography. After conversion into lactones by treatment with hydrochloric acid and drying, products running at the speed of authentic samples of mannurone and gulurone could be separated by paper chromatography. In another experiment, the uronic acids were reduced with borohydride then esterified with methyl alcohol and hydrochloric acid and reduced again, giving hexitols. The formation of D-fructose by the action of a specific D-man-nitol dehydrogenase showed the hexitols to consist largely of D-mannitol, and gas liquid chromatography of the hexaacetates of the hexitols indicated the presence of glucitol as well as mannitol.

Some guluronic acid, separated from the mixture of uronic acids by electrophoresis, was reduced by borohydride to gluconic acid, shown by a specific enzymic reaction to be D-gluconic acid (8) the expected product of reduction of the carbonyl group of L-guluronic acid (7).

$$
\begin{array}{ccc}
\left[\begin{array}{c}
\text{CHOH} \\
| \\
\text{HOCH} \\
| \\
\text{O} \quad \text{HOCH} \\
| \\
\text{HCOH} \\
| \\
\text{CH} \\
\end{array}\right] & \xrightarrow{\ \text{BH}_4\ } &
\begin{array}{c}
\text{CH}_2\text{OH} \\
| \\
\text{HOCH} \\
| \\
\text{HOCH} \\
| \\
\text{HCOH} \\
| \\
\text{HOCH} \\
\end{array}
\quad \circlearrowleft \quad
\begin{array}{c}
\text{COOH} \\
| \\
\text{HCOH} \\
| \\
\text{HOCH} \\
| \\
\text{HCOH} \\
| \\
\text{HCOH} \\
\end{array} \\
\text{COOH} & & \text{COOH} \qquad\qquad \text{CH}_2\text{OH} \\
\textbf{(7)} & & \textbf{(8)} \\
\text{L-Guluronic acid} & & \text{D-Gluconic acid}
\end{array}
$$

(a) *Synthesis of Alginic Acid*. By homogenizing *Fucus gardneri* in a medium containing polyvinylpyrrolidone, Lin and Hassid (1966b) were able to obtain extracts from which particulate fractions having various enzymic activities were obtained by centrifuging at 20,000 $\times g$ and 100,000 $\times g$. Coarse material had been removed previously by straining the homogenate and centrifuging at 1000 $\times g$ for five minutes. By the use of radioactive substrates, a series of reactions forming steps in the synthesis of polymannuronic acid from mannose were shown to be catalysed by these preparations. The reactions studied were as follows:

$$\text{D-Mannose} \xrightarrow[\text{ATP}]{\text{hexokinase}} \text{D-mannose-6-P} \xrightarrow{\text{phosphomannose mutase}} \alpha\text{-D-mannose-1-P}$$

$$\alpha\text{-D-Mannose-1-P} + \text{GTP} \xrightarrow{\text{pyrophosphorylase}} \text{GDP-D-mannose} + \text{PPi}$$

$$\text{GDP-D-mannose} + 2\text{NAD}^+ \text{ or } 2\text{NADP}^+ + \text{H}_2\text{O} \xrightarrow{\text{dehydrogenase}}$$
$$\text{GDP-D-mannuronic acid} + 2\text{NADH or } 2\text{NADPH} + 2\text{H}^+$$

$$\text{GDP-D-mannuronic acid} \xrightarrow{\text{transferase}} \text{polymannuronate}$$

ATP = adenosine triphosphate; GTP = guanosine triphosphate; PPi = inorganic phosphate; NAD = nicotinamide-adenine dinucleotide (Coenzyme I); NADP = NAD-phosphate (Coenzyme II); NADH and NADPH are reduced forms.

The first stage in the synthesis is the phosphorylation of D-mannose to D-mannose 6-phosphate, and this is the first time that phosphorylation of a hexose has been demonstrated in a cell-free extract of a marine alga. The radioactive sugar phosphate formed from ^{14}C labelled mannose had the electrophoretic mobility of a sugar phosphate, and was shown by specific enzymic reactions to be the 6-phosphate.

Guanosine diphosphate (GDP) mannose was synthesized from guanosine triphosphate (GTP) and D-mannose 1-phosphate and in poorer yield from D-mannose and GTP, indicating that both a phosphomannose mutase and a pyrophosphorylase are present in the enzyme preparation.

GDP-D-mannose dehydrogenase activity was found to be rather greater in the supernatant from centrifuging the seaweed homogenate at $100,000 \times g$ than in the separated particulate fraction and the rate of formation of radioactive GDP-D-mannuronic acid from GDP-D-mannose was increased by the presence in the solution of nicotinamide adenine dinucleotide or its phosphate. The GDP-D-mannuronic acid was characterized by methods previously described (Lin and Hassid, 1966a). GDP-D-mannuronate ^{3}H and GDP-D-mannuronate ^{14}C were used to demonstrate the incorporation of the uronic acid into alginic acid in the presence of the different enzymic fractions. Sodium alginate was added to the mixture after incubation and radioactivity was found in the precipitated alginic acid after a series of standard purification steps. Radioactive mannurone was characterized by chromatography in the hydrolysate of this alginic acid and electrophoresis indicated the presence of 4-mannosylmannitol after reduction of a radioactive dimannuronate separated by chromatography from the products of partial hydrolysis, affording evidence of 1,4-linked mannuronic acid in the product synthesized.

Alginic acid also contains L-guluronic acid residues, and as GDP-L-guluronic acid is present in the seaweed, the final stage of alginic acid synthesis probably involves a succession of transfers of uronic acid from the two GDP-uronic acids to an acceptor molecule forming β-1, 4-linked polyuronide chains.

$$\left. \begin{array}{c} \text{GDP-D-mannuronic acid} \\ \text{or} \\ \text{GDP-L-guluronic acid} \end{array} \right\} + \text{acceptor} \xrightarrow[\text{transferase}]{\text{glycosyl}} \text{alginic acid}$$

This last stage of the synthesis has not yet been achieved in the laboratory, nor has the epimerization of GDP-D-mannuronic acid to GDP-L-guluronic acid, although this conversion probably takes place in the algal cells. It is not unlikely that there is some biochemical relationship between the D-mannose shown to be a precursor of alginic acid, and mannitol found in the brown algae, but an enzyme system responsible for their interconversion in these plants has yet to be established.

In extended studies of the metabolism of *F. vesiculosus* the composition of samples was followed over a period of seventy-four hours with alternating periods of light and darkness. After an initial five-hour period of photosynthesis in which portions of frond made use of $^{14}CO_2$ (Bidwell, 1966), mannitol was the main product of photosynthesis, accounting for 85% of the total activity at the end of the photosynthetic period, but some more complex compounds, including fucoidan and alginic acid had also formed. During the subsequent period of culture without added radioactive carbon, there was a considerable increase in the amount of radioactive alginic acid and fucoidan, but the increase in the radioactivity in complex compounds was only about a tenth of the loss of that in mannitol. The rate of synthesis of alginic acid and fucoidan as measured by their radioactivity was much higher in the initial photosynthetic period than in later stages of the experiment during which the rate tended to be rather higher in the periods of darkness than in the light. It was therefore concluded that the polysaccharides are derived not directly from mannitol but from a common precursor.

In earlier work (Bidwell et al., 1958), which was limited to the photosynthetic period, it was surmised that part of the alginic acid could be an active metabolite. Although this does not now appear to be the case, the insoluble residue, which remains after extraction of the tissues with acid and alkali, seems to be actively involved in the metabolism of the plant, as its radioactivity increases in darkness and becomes less when exposed to light.

The finding that alginic acid was not formed in appreciable quantities when *F. vesiculosis* tissue was supplied with radioactive mannitol (Bidwell and Ghosh, 1962) is in agreement with the conclusions reached in the recent work that the synthesis of mannitol and alginic acid from carbon dioxide follow independent pathways. In similar experiments with other radioactive metabolities (Bidwell and Ghosh, 1963), little alginic acid was synthesized from glucose but about a third of the acetate taken up by the tissues appeared as "normal" (acid insoluble but alkali soluble) alginic acid. A rather smaller proportion was formed from pyruvic acid.

On the other hand, Lin and Hassid (1966b), using radioactive sugars, found that discs of *Fucus gardneri* were able to synthesize alginic acid from mannose in considerably greater yield than from glucose. No radioactive mannitol was detected in this experiment.

(b) *Synthesis of Laminaran.* Although the biosynthesis of laminaran has not been demonstrated with an enzyme system of brown algae, a β-1,3-glucan has been synthesized from UDP-glucose using a transferase from the fresh-water flagellate *Euglena* (Goldemberg and Marechal, 1963; Marechal and Goldemberg, 1964). The presence of UDP-glucose in a brown alga (Lin and Hassid, 1966a) suggests that this substance may be the precursor of laminaran, but the parallel seasonal variations of laminaran and mannitol may indicate that mannitol also has some biological relationship to the polyglucan. As a result of the appearance of some fructose as well as glucose, with loss of laminaran in *L. digitata* fronds held in the dark in seawater saturated with chloroform, it has been suggested (Quillet, 1954) that enzymes in the plant can lead to interconversion of mannitol and laminaran, via fructose and glucose.

Extracts of *L. digitata, Rhodymenia palmata, Cladophora rupestris* and *Ulva lactuca* have been shown to have some β-transglucosylation activity (Duncan *et al.*, 1956), and in the case of *Cladophora* and *Ulva* extracts, the synthesis of β-1,3-linkages has been demonstrated (Duncan *et al.*, 1959).

(c) *Synthesis of Fucoidan.* The greater part of the fucoidan molecule is composed of residues of sulphated fucose so that the GDP-fucose found in the extract from *F. gardneri* (Lin and Hassid, 1966a) may be the biological precursor of fucoidan. The GDP-fucose may, in turn, be derived from GDP-mannose.

Although these conversions have not been demonstrated in the algae, enzymes extracted from *Aerobacter aerogenes* have been shown (Ginsberg, 1960) to convert GDP-D-mannose into GDP-L-fucose.

Some radioactive fucose was found in a polymer synthesized by discs of *F. gardneri* in the presence of radioactive mannose or glucose, and was also present as a minor product in the hydrolysate of neutral sugar nucleotides after incubation of radioactive GDP-D-mannose with the enzyme preparations from the seaweed (Lin and Hassid, 1966b).

The incorporation of $S^{35}O_4^{11}$ into fucoidan has been demonstrated (Bidwell and Ghosh, 1963b) using *F.* vesiculosus. From the active exchange of SO_4^{11}, synthesis of the polysaccharide is thought to precede sulphation.

2. *Nucleotides in the Rhodophyceae*

Following the identification of UDP-glucose and UDP-galactose in the products of photosynthesis by *Iridaea flaccidum* (Bean and Hassid, 1955), a more complete investigation was made of the nucleotides present in the red seaweed *Porphyra perforata* (Su and Hassid, 1962). Nucleotides were precipitated from the alcohol extract of the seaweed with mercuric acetate, and the mercury salts, suspended in water, were decomposed with hydrogen sulphide. The mixed nucleotides were then adsorbed on a Dowex 1 column and fractionated by elution with 0·01N-hydrochloric acid containing increasing concentrations of sodium chloride. The nucleotides were characterized chiefly by spectrophotometric analysis, electrophoretic and chromatographic mobility and pentose and phosphate content. After mild hydrolysis, the hexoses were separated by paper chromatography and identified. By this means the presence of GDP-D-mannose GDP-L-galactose, UDP-D-glucose, UDP-D-galactose and UDP-glucuronic acid was demonstrated. The glucose and mannose nucleotides were present in considerably greater amount than the galactose derivatives. In addition, a nucleotide adenosine 3',5'-pyrophosphate was isolated for the first time.

(a) *Synthesis of Porphyran*. The polysaccharide, porphyran, which is obtained from *Porphyra* and related genera, contains sulphated residues derived from both D- and L-galactose. Su and Hassid (1962) suggest that the nucleotides found in the seaweed are involved in the synthesis of porphyran according to the following scheme:

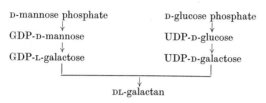

In porphyran some of the L-galactose residues are esterified with sulphate in the 6 position while the others are replaced by 3,6-anhydro-L-galactose; some of the D-galactose residues are in the form of the 6-*O*-methyl ether. The authors suggest that the adenosine 3′,5′-pyrophosphate in *P. perforata* activates inorganic sulphate involved in the enzymic sulphation of the galactose residues.

The Bangor School (Peat and Rees, 1961; Rees, 1961a) have studied the sulphatase activity of a 0·25% sodium carbonate extract of *P. umbilicalis*. Besides the sulphatase activity, the extract brought about a decrease in the viscosity of a porphyran substrate; the fragments formed were still relatively large as no increase in reducing power could be observed. The sulphatase activity of the preparation could be increased twenty-two fold by adsorbtion on calcium phosphate gel, followed by elution with sodium sulphate, but the depolymerase activity was retained. The sulphatase activity was greatest with porphyran as a substrate, and weak or doubtful with sulphated oligosaccharide and monosaccharide sulphates. The enzymic activity appears to be dependent on a bi- or ter-valent cation (shown not to be Mg^{++}) as it is inhibited by metal binding agents, but this effect may be due to the interaction of cations with the porphyran substrate. Similarly, the strong activation by borate may be due to the formation of a complex with hydroxyl groups in the polymer thus altering the molecular charge and configuration.

Using this enzyme with porphyran as substrate, it was shown (Rees, 1961b) that equimolar amounts of free sulphate and 3,6-anhydrogalactose are formed. The two reactions have similar pH characteristics, are both activated by borate, and the activities are inhibited and restored by the same reagents. It is therefore probable that the 3,6-anhydro-L-galactose units in porphyran are formed in the plant cells by the action of this enzyme on L-galactose 6-sulphate. It is suggested that the enzyme should more correctly be called a "sulphate eliminase" as it probably cleaves the C—O bond of the sulphate

2+

ester. The name sulphatase should be reserved for those enzymes which hydrolyze sulphate esters by fission of the O—S bond.

C. UTILIZATION OF POLYSACCHARIDES IN RESPIRATION

If the path of utilization of polysaccharides parallels those of the higher plants and animals, an early step in the process is the breakdown of the polysaccharides to monosaccharides or their derivatives. Although enzymes have been found which will degrade the more widely studied algal polysaccharides, the sources of the enzymes are, in most cases, animals or microorganisms (see sections on individual polysaccharides).

Extracts from several red, brown and green seaweeds, however, hydrolyse a number of polysaccharides, including algal floridean starch and laminaran, the end product being largely glucose. It was noted that fucoidan was not attacked by any of the extracts (Duncan *et al.*, 1956; Manners and Mitchell, 1967).

Enzymes from *P. umbilicalis* were found to bring about marked hydrolysis of floridean starch and laminaran as well as various glucosides examined as substrates, but apart from desulphation (see p. 21), the only action on porphyran was a decrease in viscosity of the solution. No low molecular weight fragments could be detected (Peat and Rees, 1961).

It seems probable that there are a number of ways in which energy is obtained from the oxidation of sugars, and that the metabolic pathways vary from one class of algae to another. The occurrence of intermediate metabolites and enzymes involved in different cycles is discussed by Gibbs (1962) and by Jacobi (1962).

The steady loss of radioactive mannitol from *F. vesiculosus* frond after its initial photosynthesis from $^{14}CO_2$ indicates that mannitol is the main respiratory substrate in this alga (R. G. S. Bidwell, 1966, personal communication). In earlier experiments, ^{14}C-labelled mannitol was dissolved in aerated seawater in which the alga was kept in the dark (Bidwell and Ghosh, 1962). In these conditions only a small fraction of the carbon dioxide respired came from the radioactive mannitol, probably because the non-radioactive mannitol already present in the cells was utilized preferentially. The rate of respiration of radioactive carbon dioxide increased during the thirteen hours of the experiment, suggesting that as the exogenously supplied mannitol was slowly absorbed into the tissues, it became available for respiration. At the end of the experiment the largest part of the radioactive carbon found in the tissues was present in the insoluble residue.

VII. FUNCTIONS OF POLYSACCHARIDES IN THE ALGAE

Although there is no definite proof of the part played by any polysaccharide in the life cycle of algae, the topics which have been discussed in this chapter provide circumstantial evidence for the functions of some of these compounds.

It is highly probable that starch in the green and red algae and laminaran in the brown algae are short- or long-term food reserve materials. Their hydrolysis to glucose by enzymes shown to be present in the algae, and the very general utilization of glucose for respiration by living organisms of all kinds, suggests that this is their role. The gradual disappearance of laminaran from tissues of *Laminaria* spp. during the winter when the rate of respiration may exceed that of photosynthesis points in the same direction.

Likewise, there is little doubt that the highly orientated polysaccharides, such as cellulose and some mannans and xylans, form the basic structure of the cell walls and contribute largely to the mechanical strength of the tissues. These substances and their functions are similar to those in higher plants, but in very many of the algae a large part of the weight of the plant is made up of polysaccharides whose functions are much less apparent. Among such compounds the sulphated polysaccharides are outstanding, being invariably present as far as is known in algae, and absent in land plants. Their location in the plant tissues has been stated by different workers to be in the cell wall and in intercellular spaces, but decisive evidence on this point is scanty. There seems little doubt, however, that whatever their exact location they help to build up flexible structures suited to their environment. When completely submerged, the weight of a seaweed is largely counterbalanced by the buoyancy of the seawater (helped by air bladders in some species), so that a structure built up of gels or highly viscous fluids inside membranes can remain extended through a large volume of water to make the most use of available light and nutrients, while remaining flexible enough to withstand the movement of the water without damage. It is possible that in some species there has been some selection so that stiffer gels are present in plants exposed to more severe wave action (Rees and Conway, 1962).

The fact that these polysaccharides, and also alginic acid present in the Phaeophyceae, are anionic polyelectrolytes with marked cation exchange properties must be of some significance in plants living in a saline medium. For example, the sulphated polysaccharide porphyran is probably responsible for the selective absorbtion of potassium and exclusion of sodium in *P. perforata* (Eppley, 1958).

A further role of these polysaccharides may be the prevention of desiccation in plants exposed by the tides; fucoidan appears to be particularly efficient in this respect as among the Fucales those growing highest on the shores which are exposed to the atmosphere for the longest periods have on the average considerably more fucoidan (*Pelvetia canaliculata*, 18 to 24% fucoidan, *Fucus spiralis*, 18 to 22%) than those which are exposed for only short periods (*F. serratus*, 8 to 13%).

REFERENCES

Baardseth, E. (1966). *Proc. 5th int. Seaweed Symp.* (1965), Halifax, Nova Scotia (E. G. Young and J. L. McLachlan, eds), p. 19, Pergamon Press, Oxford.
Bean, R. C., and Hassid, W. Z. (1955). *J. biol. Chem.* **212**, 411.
Biebl, R. (1962). *In* "Physiology and Biochemistry of the Algae" (R. A. Lewin, ed.), p. 799, Academic Press, New York and London.
Bidwell, R. G. S. (1957). *Can. J. Bot.* **35**, 945.
Bidwell, R. G. S. (1958). *Can. J. Bot.* **36**, 337.
Bidwell, R. G. S. (1966). Personal communication.
Bidwell, R. G. S., and Ghosh, N. R. (1962). *Can. J. Bot.* **40**, 803.
Bidwell, R. G. S., and Ghosh, N. R. (1963). *Can. J. Bot.* **41**, (a) 155; (b) 209.
Bidwell, R. G. S., Craigie, J. S., and Krotkov, G. (1958). *Can. J. Bot.* **36**, 581.
Black, W. A. P. (1948). *J. Soc. chem. Ind., Lond.* **67**, 355.
Black, W. A. P. (1949). *J. Soc. chem. Ind., Lond.* **68**, 183.
Black, W. A. P. (1950a). *J. mar. biol. Ass. U.K.* **29**, 45 and ref. cited therein.
Black, W. A. P. (1950b). *J. Soc. chem. Ind., Lond.* **69**, 161.
Black, W. A. P. (1954a). *J. mar. biol. Ass. U.K.* **33**, 49.
Black, W. A. P. (1954b). *J. Sci. Fd. Agric.* **5**, 445.
Black, W. A. P., and Dewar, E. T. (1949). *J. mar. biol. Ass. U.K.* **28**, 673.
Black, W. A. P., Richardson, W. D., and Walker, F. T. (1959). *Econ. Proc. R. Dubl. Soc.* **4**, 137.
Black, W. A. P., Blakemore, W. R., Colquhoun, J. A., and Dewar, E. T. (1965). *J. Sci. Fd. Agric.* **16**, 574.
Brody, M., and Brody, S. S. (1962). *In* "Physiology and Biochemistry of Algae" (R. A. Lewin, ed.), p. 3, Academic Press, New York and London.
Clendinning, K. A. (1964). *Proc. 4th Int. Seaweed Symp.* (1961), Biarritz, (A. D. DeVirville and J. Feldman, eds), p. 55, Pergamon Press, Oxford.
Colin, H., and Guéguen, E. (1930). *C. r. hebd. Séanc. Acad. Sci., Paris* **191**, 163.
Colin, H., and Augier, J. (1939). *C. r. hebd. Séanc. Acad. Sci., Paris* **208**, 1450.
Craigie, J. S., McLachlan, J, Majak, W., Ackman, R. G., and Tocher, C. S. (1966). *Can. J. Bot.* **44**, 1247.
Cronshaw, J., Myers, A., and Preston, R. D. (1958). *Biochem. biophys. Acta* **27**, 89.
Duncan, W. A. M., Manners, D. J., and Ross, A. G. (1956). *Biochem. J.* **63**, 44.
Duncan, W. A. M., Manners, D. J., and Thompson, J. L. (1959). *Biochem. J.* **73**, 295.
Eppley, R. W. (1958). *J. gen. Physiol.* **41**, 901.
Falk, M., Smith, D. G., McLachlan, J., and McInnes, A. G. (1966). *Can. J. Chem.*, **44**, 2269.
Fanshawe, R. S., and Percival, Elizabeth (1958). *J. Soc. Fd. Agric.* **9**, 241.
Fischer, F. G., and Dörfel, H. (1955). *Hoppe-Seyler's Z. physiol. Chem.* **302**, 186.
Ford, C. W., and Percival, Elizabeth (1965). *J. chem. Soc.* (a) p. 7035; (b) p. 7024.
Frei, E., and Preston, R. D. (1961). *Nature, Lond.* **192**, 939.
Frei, E., and Preston, R. D. (1964). *Proc. roy. Soc.* **160**B, (a) p. 293; (b) p. 314.
Gibbs, M. (1962). *In* "Physiology and Biochemistry of Algae" (R. A. Lewin, ed.), p. 61, Academic Press, New York and London.
Ginsberg, V. (1960). *J. biol. Chem.* **235**, 2196.
Ginsberg, V. (1964). *In* "Advances in Enzymology" (F. F. Nord, ed.), Vol. **26**, 35, Interscience, New York, London and Sydney.
Goldemberg, S. H., and Marechal, L. R. (1963). *Biochem. biophys. Acta* **71**, 743.
Harvey, H. W. (1955). *In* "Chemistry and Fertility of Seawater". Cambridge University Press, London.
Iriki, Y., and Miwa, T. (1960). *Nature, Lond.* **185**, 178.
Iriki, Y., Suzuki, T., Nisizawa, K., and Miwa, T. (1960). *Nature, Lond.* **187**, 82.

Jacobi, G. (1962). *In* "Physiology and Biochemistry of Algae" (R. A. Lewin, ed.), p. 125, Academic Press, New York and London.

Jones, R. F. (1956). *Biol. Bull.* **110**, 169.

Levring, R. (1947). *Göteborgs K. Vetensk. o. Vitterh. Samh. Handl.* Ser. B5, p. 1.

Lin, T. Y., and Hassid, W. Z., (1964). *J. biol. Chem.* **234**, PC944.

Lin, T. Y., and Hassid, W. Z. (1966a). *J. biol. Chem.* **241**, 3282.

Lin, T. Y., and Hassid, W. Z. (1966b). *J. biol. Chem.* **241**, 5284.

Lindberg, B. (1955a). *Proc. 2nd int. Seaweed Symp.* (T. Braarud and N. A. Sorensen, eds), p. 33, Pergamon Press, London, New York and Paris.

Lindberg, B. (1955b). *Acta chem. scand.* **9**, 1093.

Lindberg, B. (1955c). *Acta chem. scand.* **9**, 1097.

Lindberg, B., and Paju, J. (1954). *Acta chem. scand.* **8**, 817.

Love, J., and Percival, Elizabeth (1964a). *J. chem. Soc.* p. 3345.

Love, J., and Percival, Elizabeth (1964b). *J. chem. Soc.* p. 3338.

McCully, M. E. (1965). *Can. J. Bot.* **43**, 1001.

Mackie, I. M., and Percival, Elizabeth (1959). *J. chem. Soc.* p. 1151.

McKinnell, J. P., and Percival, Elizabeth (1962). *J. chem. Soc.* p. 3141.

Macpherson, M. G., and Young, E. G. (1952). *Can. J. Bot.* **30**, 67.

Majak, W., Craigie, J. S., and McLachlan, J. L. (1966). *Can. J. Bot.* **44**, 541.

Manners, D. J., and Mitchell, J. P. (1967). *Proc. Biochem. Soc.* March, p. 15.

Marechal, L. R., and Goldemberg, S. H. (1964). *J. biol. Chem.* **239**, 3163.

Meeuse, B. J. D. (1962). *In* "Physiology and Biochemistry of the Algae" (R. A. Lewin, ed.), p. 300, Academic Press, New York and London.

Meeuse, B. J. D., Andries, M., and Wood, J. A. (1960). *J. exp. Bot.* **11**, 129.

Moss, B. L. (1948). *Ann. Bot.* **12**, 267.

Neufeld, E. F., and Hassid, W. Z. (1963). *In* "Advances in Carbohydrate Chemistry" (R. L. Whistler and J. N. BeMiller, eds), Vol. **18**, p. 309, Academic Press, New York and London.

Parke, M., and Dixon, P. S. (1964). *J. mar. biol. Ass. U.K.* **44**, 499.

Peat, S., and Rees, D. A. (1961). *Biochem. J.*, **79**, 7.

Peat, S., Turvey, J. R., and Rees, D. A. (1961). *J. chem. Soc.* p. 1590.

Preston, R. D. (1965). *Advt. Sci., Lond.* **22**, 1.

Putman, E. W., and Hassid, W. Z. (1954). *J. Am. chem. Soc.* **76**, 2221.

Quillet, M. (1954). *C. r. hebd. Séanc. Acad. Sci., Paris*, **238**, 926.

Quillet, M. (1958). *C. r. hebd. Séanc. Acad. Sci., Paris*, **246**, 812.

Rees, D. A. (1961a). *Biochem. J.* **80**, 449.

Rees, D. A. (1961b). *Biochem. J.* **81**, 347.

Rees, D. A., and Conway, E. (1962). *Biochem. J.* **84**, 411.

Silva, P. S. (1962). *In* "Physiology and Biochemistry of Algae" (R. A. Lewin, ed.), p. 827, Academic Press, New York and London.

Smith, G. M. (1955a). "Cryptogamic Botany" **1**, 2nd Ed., p. 293, McGraw-Hill, London.

Smith, G. M. (1955b). "Cryptogamic Botany" **1**, 2nd Ed., p. 15, McGraw-Hill, London.

Stoloff, L., and Silva, P. (1957). *Econ. Bot.* **11**, 327.

Strain, H. H. (1951). *In* "Manual of Phycology" (G. M. Smith, ed.), p. 243, Chronica Botanica Co., Waltham, Mass., U.S.A.

Su, J. C., and Hassid, W. Z. (1962). *Biochemistry, N.Y.* **1**, 474.

Takeda, H., Nisizawa, K., and Miwa, T. Private communicaton

Thiele, H., and Andersen, G. (1955). *Kolloidzeitschrift* **143**, 21.

Young, E. Gordon (1966). *Proc. 5th int. Seaweed Symp.* (1965), Halifax, Nova Scotia (E. G. Young and J. L. McLachlan, eds), p. 337, Pergamon Press, Oxford.

CHAPTER 2

The Elucidation of the Structure
of Polysaccharides[1]

I. INTRODUCTION

Before the structure of any polysaccharide can be determined, it is necessary in the first instance to make certain that the polymer has been isolated from a single specie. The collection of pure species is particularly difficult with some genera, for example *Enteromorpha*, of marine algae, and the services of a trained algologist are often necessary. Secondly, most extraction methods yield mixtures of polysaccharides contaminated with other substances such as protein. At the same time it should be borne in mind that extraction procedures may modify the actual structure of the molecule and also alter its molecular weight distribution. For these reasons the extraction conditions should be as mild as possible. There is, however, no standard method for the isolation of algal polysaccharides. Each class of seaweed presents its own problems and the task of separation and purification is often difficult and tedious, but is an essential preliminary to structural investigations.

Standard techniques for the removal of protein (Fisher and Percival, 1957; Larsen *et al.*, 1966) can be applied to these materials, but in some instances complete removal has proved impossible and a covalent link between the protein and the polysaccharide seems very probable.

II. ISOLATION OF PURE POLYSACCHARIDE MATERIAL

A. DIFFERENTIAL EXTRACTION

Some separation of the polysaccharides in an alga is often possible by differential extraction. For example, a preliminary separation of polysaccharides can be made by successive extraction with solvents of increasing power. The sequence, cold water, hot water, cold dilute alkali and hot dilute

---

[1] The nomenclature used throughout the monograph is that recommended in the Handbook for Chemical Society Authors (1960) Special publication No. 14, p. 140, as modified by Rules of Polysaccharide Nomenclature, Proposed revisions, May 31st, 1966 by the Committee on Carbohydrate Nomenclature, American Chemical Society, and IUPAC-IUB recommendations *J. biol. Chem.* 1966, **241**, 527.

alkali, will yield fractions of differing composition and properties. This is true for the extraction of *Laminaria* spp.; water removes laminaran contaminated with small quantities of fucoidan. Treatment of the residual weed with dilute acid yields mainly fucoidan, and extraction of the weed residue with sodium carbonate dissolves out the polyuronide, alginic acid. The residual material still comprises other uronic acid-containing polysaccharides and a small amount of cellulose. On the other hand, aqueous extraction of the green sea-weed, *Codium fragile*, yields a mixture of a sulphated arabinogalactan, a starch, and other polysaccharides together with protein (Love and Percival, 1964a). Subsequent mild chlorite treatment of the weed and dilute alkali extraction gives a mannan contaminated with silica, but treatment of the weed residue with 20% alkali leads to the separation of a pure mannan from the last solution.

B. FRACTIONAL PRECIPITATION

There are no standard methods for the separation of mixtures of poly-saccharides, as again each biological material presents its own specific diffi-culties. Fractionation of seaweed mucilages into individual polysaccharides has been mainly achieved so far by fractional precipitation in organic solvents or by selective precipitation from aqueous solutions by certain electrolytes. The gradual addition of ethanol to aqueous solutions of water-soluble poly-saccharides has been widely practised outside the algal field, although on the whole the separation is poor. However, fucoidan (see p. 158) and λ-carrageenan (see p. 146) have, by this means, been separated into fractions of different composition.

The use of metallic ions as precipitating agents has proved more satisfac-tory. The addition of dilute potassium chloride to an aqueous solution of carrageenan precipitates κ-carrageenan and leaves λ-carrageenan in solution (Smith and Cook, 1953). Separation of alginic acid into mannurone- and gulurone-rich fractions can be brought about in the same way (Haug and Smidsrod, 1965), or by the addition of manganous salts (McDowell, 1958).

Precipitation by a specific complexing agent is a method of general appli-cation. The most widely known example of this technique is perhaps the al-most complete separation of an amylose complex from a starch solution on addition of a polar organic molecule (Schoch, 1945). Metallic salts have been frequently used as specific precipitating agents for water-soluble polysac-charides; the formation of copper complexes with Fehling's solution, copper chloride, copper sulphate, and copper acetate has been used. An excess of precipitant is added and the insoluble "polysaccharide-copper complex" is decomposed by an alcoholic solution of acid or by a chelating agent. The mannans from *Porphyra umbilicalis* (Jones, 1950), and from *Codium fragile*

(Love and Percival, 1964b), were isolated by complexing with Fehling's solution. Barium hydroxide is another complexing agent which has proved of value (Meier, 1958), and gave a partial fractionation of the water-soluble polysaccharides of *Caulerpa filiformis* (Mackie and Percival, 1961).

Another type of complexing agent, especially useful for charged polysaccharides, includes such quaternary ammonium salts as cetyltrimethylammonium (CTA) bromide and cetylpyridinium (CP) bromide which form with polyanions salts that are very insoluble in water. They can therefore be utilized to separate neutral from acidic polysaccharides (Scott, 1960). For example, the acidic fraction agaropectin, found in agar, can be separated from the neutral agarose by precipitation with cetylpyridinium chloride (Hjerten, 1962). In a similar manner, a starch-type polysaccharide was separated from the sulphated water-soluble mucilage extracted from *C. filiformis* (Mackie and Percival, 1960) by cetyltrimethylammonium hydroxide in the presence of borate. In contrast, this reagent gave no fractionation of the mucilage from *Acrosiphonia arcta* (O'Donnell and Percival, 1959b).

C. Fractionation by Column Chromatography

The introduction of cellulose ion-exchangers such as diethylaminoethyl (DEAE)-cellulose for column chromatography made available new adsorbents especially suited for the fractionation of water-soluble high molecular weight polysaccharides. Acidic polysaccharides are readily adsorbed on this material at pH values near 6, and depending on their content of acidic groups, elution has been achieved by increasing concentration of (*a*) salts, (*b*) alkaline solutions and (*c*) acidic solutions. Neutral polysaccharides are not usually, or only weakly, retained on the column at pH 5–6.

In this way the sulphated water-soluble mucilage from *Codium fragile* was separated into four fractions of increasing sulphate content by gradient elution with potassium chloride (Love and Percival, 1964a). A similar fractionation of the mucilage from *Cladophora rupestris* resulted both on DEAE-cellulose and DEAE-sephadex (Hirst *et al.*, 1965). In all these experiments with sulphated polysaccharides, a considerable proportion of the material was irreversibly bound on the cellulose. In contrast a similar fractionation of alginic acid failed to separate mannurone-rich and gulurone-rich material, but yielded two distinct fractions of similar composition but different viscosity (Hirst *et al.*, 1964).

D. Use of Enzymes

It is possible in some cases to use enzymes to degrade, preferentially, contaminating material in a polysaccharide mixture. For example, papain can

remove protein and α-amylase will remove starch-type polysaccharide. However, if crude enzyme preparations are used, there is a danger of the structure
of the residual polysaccharide being modified. For example, the use of pro-
teolytic enzymes or carbohydrases should be avoided in the isolation of
starches since α-amylase is a trace contaminant in most enzyme preparations.

III. ELUCIDATION OF THE FINE STRUCTURE

Polysaccharides are high condensation polymers of monosaccharides which
are formed by the elimination of a molecule of water between each pair of
monosaccharide units

$$nC_6H_{12}O_6 \rightarrow (C_6H_{10}O_5)n + nH_2O$$

It must be emphasized that the linkage is always glycosidic, that is, through
the C-1 of an aldose sugar (C-2 in ketoses) to any of the remaining hydroxyl
groups of a second sugar, resulting in long chain molecules which may or
may not be branched.

α-Glycosidic link
in D-aldose

β-Glycosidic link
in D-ketose

In the determination of the fine structure of a polysaccharide, it is necessary to discover the following facts:

(a) The constitutent monosaccharides which make up the macromolecule.
(b) Which hydroxyl groups in addition to the glycosidic group in these
    monosaccharides are involved in linkage in the polysaccharide (linkage
    analysis).
(c) Whether the glycosidic linkage is α or β.
(d) When more than a single monosaccharide is present, the relative posi-
    tions of the different units in the macromolecule.
(e) When present, the site of ester sulphate groups.
(f) The average chain length, and in branched molecules the position of the
    branches and the degree of branching.
(g) The molecular weight and the overall shape of the molecule.

Few of the methods available provide unequivocal information on any of
these points and it is desirable to confirm the findings by all the available
techniques.

2*

Many algal polysaccharides present additional problems. Uronic acid residues and sulphate ester groups are frequently present, and these interfere with many of the standard methods commonly used in structural determinations. Nevertheless, many of the techniques in use for the elucidation of polysaccharide structure have been applied to algal polysaccharides. It is therefore appropriate at this stage, in order to avoid frequent repetition, to describe these, and also the special modifications introduced by reason of the presence of the ionic groups so that the reader will be familiar not only with the method but with the interpretation of the results when applied to a particular polysaccharide. In addition the general methods used for the determination of the site of the sulphate groups will be discussed.

A. Analysis of the Constituent Sugars
(Kowkabany, 1954; Binkley, 1955)

The first step in the determination of polysaccharide structure is the elucidation of the constituent sugars. This is done by hydrolysing about 20–50 mg of the polysaccharide with acid, and analysing the resulting syrup on a paper chromatogram. Since there is always some degradation of the resulting monosaccharides, the hydrolysis should be carried out under the mildest possible conditions for the particular polysaccharide. When the constituent sugars have been tentatively identified by the colour and mobility of their spots, a second paper is eluted on which the authentic sugars are run alongside the polysaccharide hydrolysate. This must never be taken as proof of the presence of a particular sugar, since different substances may have the same chromatographic mobility and colour of spot, as is indeed the case with mannuronic and guluronic acids and with fucose and 6-O-methylgalactose. If sufficient material is available, the sugars should be separated on a cellulose or charcoal column, or in the case of uronic acids, on a resin column and characterized as crystalline materials and derivatives.

Hydrolysis to the monouronic acid is sometimes impossible (O'Donnell and Percival, 1959b), due to the drastic conditions necessary to cleave the glycosidic links degrading the uronic acid as it is liberated. However the uronic acid content of a polysaccharide can be found by measurement of the carbon dioxide released on treatment with 19% hydrochloric acid (Nevell, 1963), but pentoses also yield a small amount of carbon dioxide under these conditions and results of 5% or less uronic acid, if determined in the presence of pentans, should be interpreted with caution. Colorimetric determinations (Brown and Hayes, 1952; Jones and Pridham, 1953; Kaye and Kent, 1953; Gregory, 1960) of uronic acid are also subject to the same limitations since, at the best, they are only accurate to within $\pm 2\%$, and other acids such as acetic acid derived from acetyl residues interfere.

Where the supply of material is too small for crystallization, paper chromatography may be supplemented by paper electrophoresis (= ionophoresis) and gas liquid chromatography of the trimethylsilane derivatives. Although sugars themselves are neutral substances and do not move on a paper ionophoretogram under the influence of an electric current, they have the property through their polyhydroxy structure of forming ionizing complexes with borate, molybdate and tungstate (Weigel, 1963). It has been found possible to separate mixtures of sugars by electrophoresis, in for example, borate buffer. This has the advantage that the sugars have different relative mobilities from those found by ordinary paper chromatography. The mixture of gulose and mannose derived from reduced alginic acid (see p. 103) gave a single spot on paper chromatograms eluted with many different solvents, but gave two distinct spots when subjected to ionophoresis in borate buffer.

It is necessary to substitute the hydroxyl groups in order to render sugars sufficiently volatile for separation by gas liquid chromatography (Bishop, 1964). Mention will be made of the characterization of methylated sugars by this method. Acetylated derivatives have also been used, but the fully substituted trimethylsilane derivatives have the added advantages that they can be formed by simple mixing of the sugar or mixture of sugars with the reagents and applied to the column without further purification (Sweeley et al., 1963; Yamakawa and Ueta, 1964). Furthermore the free sugars can readily be recovered by the addition of water to the trimethylsilane derivatives. The use of these derivatives has proved of considerable value in the analysis of an algal polysaccharide (Elizabeth Percival, unpublished results).

B. LINKAGE ANALYSIS FROM METHYLATION STUDIES

Although the application of chromatography has brought about radical changes in the determination of polysaccharide structure, the classical technique of methylation still plays a very important role in the determination of the linkage between the different units in the polysaccharide. The procedure involves the preparation of a completely methylated polysaccharide. This often entails exhaustive treatment with the methylating agents such as dimethyl sulphate and sodium hydroxide, followed in many cases by repeated treatment with silver oxide and methyl iodide. Hydrolysis of the methylated product gives a mixture of methylated monosaccharides which may be separated and characterized. The original points of linkage will correspond to the unsubstituted hydroxyl groups in the methylated monomeric units. For example, if 2,3,6-tri-O-methylglucose is separated from the products of methylation and hydrolysis, and the stability of the polysaccharide indicates the absence of furanose sugars, it follows that these units were linked by C-1 and C-4 in the macromolecule.

Whereas some thirty-five years ago it was necessary to methylate about 10 g of polysaccharide and separate the derived glycosides by fractional distillation, today microgram quantities of these derivatives can be separated and identified by gas liquid chromatography (Aspinall, 1963). In addition, improvements in the methylation techniques (Hirst and Percival, 1965) have made possible the methylation of milligram quantities of material, and the derived methylated sugars can be separated on cellulose or charcoal columns or on paper, and characterized as crystalline derivatives.

It is sometimes possible, from the overall results of methylation, to devise a repeating unit for the macromolecule, which is the simplest structure consistent with the products. It must be emphasized, however, that this is only a "possible" structure and often alternative structures could give the same results.

Wherever possible, complete methylation is essential since undermethylation gives rise to methyl ethers which would incorrectly be interpreted to have structural significance.

C. Linkage Analysis and Relative Position of the Different Sugars by Partial Hydrolysis

In recent years, partial hydrolysis has played a very important part in structural studies on polysaccharides. Paper and column chromatography have enabled the separation of small quantities of di- and oligo-saccharides; the improvement in methylation techniques already mentioned, and the application of periodate oxidation studies, have facilitated the characterization of milligram quantities of these fragments. Not only has this confirmed the linkages between the units revealed by methylation studies on the whole polysaccharide, but it has also provided evidence as to the relative position of the different sugars in heteropolysaccharides such as the mucilages from *Ulva* (p. 180), from *Cladophora* (p. 167) and from *Porphyran* (p. 135).

D. Difficulties and Advantages Associated with Uronic Acid Residues

Uronic acid residues are frequently present in algal polysaccharides, and structural investigation by the methods of partial hydrolysis and of methylation, followed by hydrolysis of the methylated material, are hindered by the instability of the uronic acids towards acids and also by the greatly increased stability of the glycosidic link of the uranosyl residue. Furthermore, the presence of these charged groups hinders complete methylation of the polysaccharide. These difficulties have been particularly evident in the investigations on alginic acid. On the other hand, where uronic acid and neutral sugars occur in a single molecule as in the mucilage from *Ulva lactuca*, the increased stability of the uranosyl linkage has permitted the isolation and characterization of the aldobiouronic acid, 4-O-β-D-glucuronosyl-L-rhamnose, after acid hydrolysis of the polysaccharide.

The difficulties caused by the presence of uronic acid units in polysaccharides may be overcome by reduction to a neutral polysaccharide. Reduction may be effected in aqueous solution using sodium or potassium borohydride (Jones and Perry, 1957), if the acid groups are first esterified under mild conditions. Methyl esters may be formed by treatment of the acidic polysaccharide with diazomethane under conditions in which accompanying etherification is negligible. It is usually necessary to repeat the cycle of operations in order to achieve complete reduction of the uronic acid residues. Rees and Samuel (1965) recommend the formation of the trimethylsilane derivative of the methyl ester of alginic acid and reduction with lithium aluminium hydride, or preferably reduction with lithium borohydride of methyl di-O-acetylalginate.

Esterification of the mucilage from *Ulva lactuca* with methanolic hydrogen chloride and reduction with borohydride gave a nearly neutral polysaccharide after two treatments (Haq and Percival, 1966). In contrast, application of this procedure to alginic acid failed to yield a neutral polysaccharide in spite of numerous repetitions of the experiments. An alternative procedure using diborane (Brown and Subba Rao, 1960), a reagent which, unlike sodium and potassium borohydride, reduces carboxyl groups, gave a reduced alginic acid containing less than 10% of uronic acid after two treatments (Hirst *et al.*, 1964). In this method the propionated alginic acid was treated with diborane in *bis* (2-methoxylethyl)-ether, "diglyme".

E. PERIODATE OXIDATION STUDIES (Bouveng and Lindberg, 1960)

Treatment of glycol groups with periodic acid and its salts results in cleavage of the carbon chain and the formation of two aldehydic groups, one molecular proportion of periodate being reduced. Where three adjacent hydroxyl groups occur, a double cleavage of the carbon chain occurs with the formation of two aldehydic groups, the reduction of two molecular proportions of periodate and the liberation of one molecular proportion of formic acid. In general, open chain glycols are most readily oxidized, these are followed by cyclic *cis*-glycols; cyclic *trans*-glycols are more slowly oxidized.

1. Linkage Analysis

Periodate oxidation studies have yielded valuable information on algal polysaccharide structure. Hexopyranose residues linked through C-1 and C-4

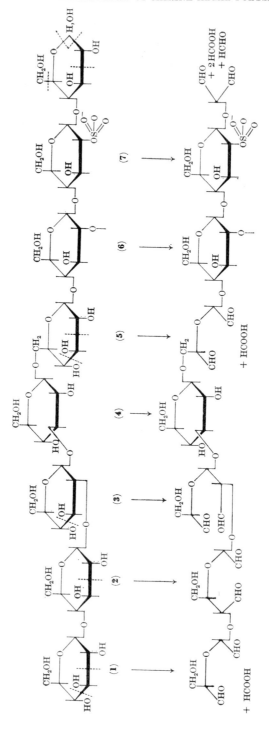

Fig. 1. The action of periodate on differently linked hexoses.

[see Fig. 1, (2)] or C-1 and C-2 (3) will reduce one mole of periodate for every unit and will themselves be cleaved through C-2 and C-3, C-3 and C-4, respectively. Non-reducing end groups (1) and non-terminal hexose units linked solely through C-1 and C-6 (5) will be attacked between C-2 and C-3, C-3 and C-4 with the reduction of two moles of periodate, and the release of one mole of formic acid per residue. Pyranose sugars linked through C-1 and C-3 (4) or units involved in branching at C-2 and C-4 (6) are immune to periodate oxidation as are carbon atoms linked to ester sulphate (7) (see Turvey, 1965). Thus, oxidation of a polysaccharide and quantitative determination of the periodate consumed, the formic acid released and the proportion of surviving sugar units (see Hay et al., 1965a) will give information concerning the nature and proportion of the glycosidic linkages present in the polysaccharide.

2. *Determination of the Degree of Polymerization*

(I) In the absence of 1,6-linked units, the proportion of formic acid produced may also be used to determine the average degree of polymerization (DP) of a linear polysaccharide. In 1,2- 1,3- and 1,4-linked polymers, three molecular proportions of formic acid are liberated per linear chain, one from the non-reducing end and two from the reducing end (Fig. 1) (see also pp. 36–39).

Similarly with branched polysaccharides, the formic acid released is a measure of the ratio of terminal to non-terminal sugar residues in the average repeating unit. In the case of a highly branched polymer, for example amylopectin, the formic acid produced from the reducing end becomes insignificant.

(II) The DP may also be determined by an assay for the reducing end-group by periodate oxidation after reduction (Hay et al., 1965c). Reduction of the polysaccharide with sodium borohydride converts the reducing terminal aldose residue into an alditol [see Fig. 2 (9, 12)] which in the case of a C-2, C-5 or C-6 linked residue (8), yields on periodate oxidation one molecular proportion of formaldehyde (10). With similar treatment, however, two molecular proportions of formaldehyde (13) are derived from a C-3 or C-4 linked residue (11).

The formaldehyde may be determined colorimetrically with chromotropic acid, after removal of the polyaldehyde and the periodate and iodate ions. It is advisable to carry out control experiments with substances of established similar constitution since the theoretical amount of formaldehyde is not always liberated. This was found to be the case with the reduced G-chains in laminaran (see p. 63) where the derived 3-linked sorbitol under the standard conditions of oxidation does not yield the expected two molecular proportions of formaldehyde.

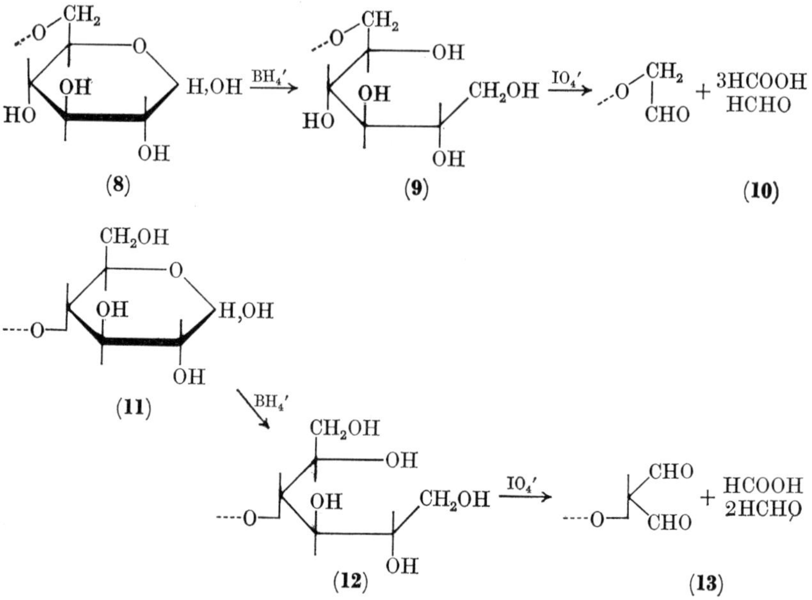

FIG. 2. Periodate oxidation of the reduced polysaccharide.

3. Linkage Analysis and the Degree of Branching

The polyaldehyde (oxopolysaccharide) obtained by periodate oxidation undergoes extensive decomposition on acid hydrolysis. If the aldehydic groups are reduced to primary alcoholic groups, with, for example boro-hydride, the resulting polyalcohol is readily hydrolysed and gives a quantitative yield of the hydrolysis products (Hay *et al.*, 1965b) (Fig. 3). Polyalcohols from 1,4-linked hexans (**14**) yield glycerol (glyceritol) (**17**) and glycollic aldehyde (**18**) from the non-reducing end, only glycerol from the reducing end and glycollic aldehyde and erythritol (**19**) or threitol from the interchain residues.

Quantitative analysis of the erythritol and glycerol in the mixture determines the proportion of terminal groups in the molecule. 6-Substituted hexose residues (**15**) give glycerol and glycolaldehyde, and 2-substituted units (**16**) yield glycerol (**17**) and glycerose (**20**). Sugars which escape oxidation due to the absence of α-glycol groups may also be detected in the hydrolysate.

4. "Overoxidation"

The reaction conditions for the periodate oxidation of polysaccharides and oligosaccharides have a profound effect on the nature and quantities of the end products formed, and in the methods mentioned above, the results are dependent on the use of the conditions detailed in the cited references.

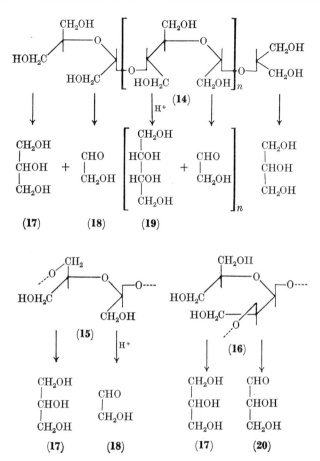

FIG. 3. Hydrolysis of polyalcohols.

Periodate oxidation of the reducing aldohexose end group of a polysaccharide [see Fig. 4 (21)], in which this is attached to the penultimate unit by a glycosidic linkage to C-4, proceeds by way of a formyl ester (22) which, after hydrolysis to a 2-O-substituted tetrose (23) is cleaved to formaldehyde and a malondialdehyde derivative (24). The latter is further oxidized by a process generally termed "overoxidation". Activation of the C—H bond adjacent to the two carbonyl groups in the malondialdehyde derivative causes further oxidation to the corresponding hydroxyl derivative (25), an acetal that can break down either by oxidation to a glyoxylic ester (26) or by hydrolysis to mesoxaldehyde (27) yielding in either case a mole of carbon dioxide and exposing the next unit to further oxidation.

A similar degradation can occur with linear polymers of aldohexopyranosyl

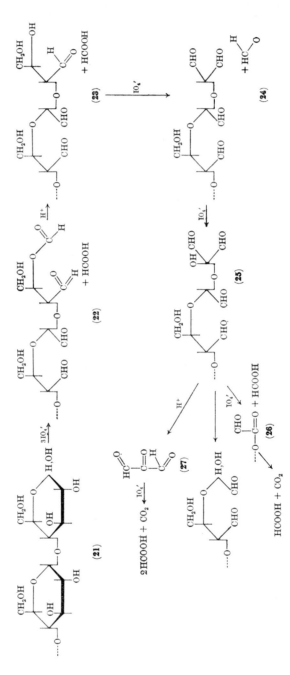

Fig. 4. "Overoxidation" of a 1,4-linked glucan.

units containing only $(1 \rightarrow 2)$ or $(1 \rightarrow 3)$ glycosidic linkages and any of these three types of polysaccharide can be completely eroded from the reducing end with the formation of one mole of formaldehyde and one mole of carbon dioxide per hexose residue.

When the reducing end group is linked through C-6 of a hexose (28) or C-5 of a pentose, periodate oxidation leads to the formation of a stable 2-O-substituted glycol aldehyde derivative (29) and "overoxidation" cannot occur.

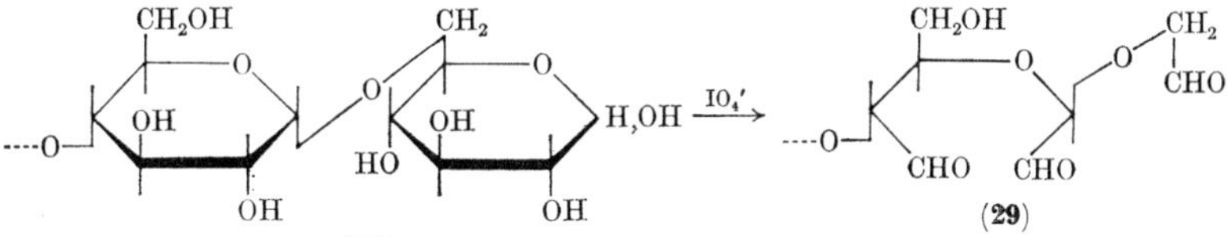

FIG. 5. Periodate oxidation of a 1,6-linked hexose.

The presence of a $(1 \rightarrow 6)$ branch linkage within a chain of aldohexosyl units joined by $(1 \rightarrow 2)$, $(1 \rightarrow 3)$ or $(1 \rightarrow 4)$ linkages, therefore, stops the oxidative erosion process and a quantitative estimate of the formaldehyde released on periodate oxidation at pH 8 reveals the position of the 1,6-linkage with respect to the reducing end unit.

The rate and extent of "overoxidation" is dependent upon various factors; the concentrations of periodate and carbohydrate, the temperature and especially the pH of the reaction mixture (Cantley et al., 1959). The hydrolysis of the intermediary formyl esters (22) is usually very slow at about pH 3·6 in 0·015M-periodate. At higher and lower pH values, the rate of hydrolysis is in-increased, being very rapid at pH 7. Increased temperature has a similar but less marked effect. The rate of oxidation can be controlled either by using the less soluble potassium periodate or by carrying out the reaction at 0° and in subdued light. Hence by carefully regulating the pH and carrying out the oxidation in dilute solution at 0°, "overoxidation" can be inhibited.

5. Smith Degradation (Goldstein et al., 1965)

Because of the marked difference in stability between true acetals and glycosides, it is possible by mild acid hydrolysis to cleave the acetal linkages in polyalcohols, resulting from periodate oxidation and reduction of polysaccharides, and to leave any glycosidic linkages intact. Those units which have been cleaved by periodate are true acetals and very sensitive to acid, whereas any unit which has not been oxidized by the periodate and is joined

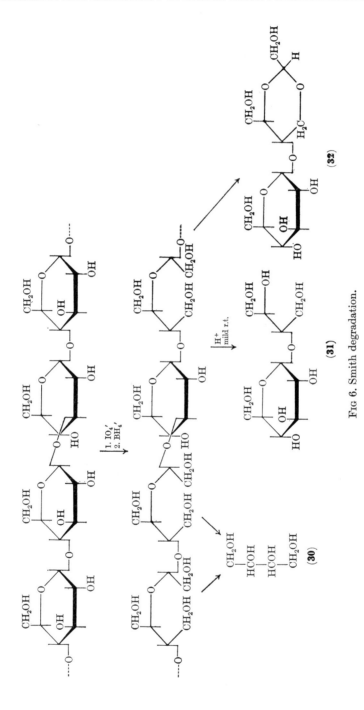

FIG 6. Smith degradation.

to a cleaved unit appears as a glycoside which is relatively stable to acid. For example, a 1,3-linked glucose unit adjacent to oxidized residues linked through the 4 or 6 position is split off as a glycoside of erythritol (31) (Fig. 6).

The formation of erythritol (30) at the same time as compound (31) indicates the presence of two or more adjacent (1 → 4) linked D-glucose units in the polysaccharide. The glycolaldehyde acetal of 2-O-β-D-glucopyranosyl-D-erythritol (32) may also be formed (Lewis and Smith, 1963), and this must be taken into account when looking for the 2-O-β-D-glucopyranosyl-D-erythritol (31).

Similarly, pentans, for example, the *Rhodymenia xylan* (see p. 89), yields xylosyl-D-glycerol from adjacent (1 → 3), (1 → 4)-linked units. Application of this technique to the polysaccharide from *Ulva* (see p. 181) has yielded valuable information on the relative position of different units and different types of linkage in the macromolecule.

6. *Barry Degradation* (Barry, 1943)

The condensation of oxopolysaccharides with carbonyl reagents such as phenylhydrazine, isonicotinhydrazide and thiosemicarbazide, and the determination of the nitrogen and sulphur contents of the condensation products enabled Barry and his colleagues to estimate the proportion of α-glycol groups present in the polysaccharide (Fig. 7) (Bouveng and Lindberg, 1960; O'Colla, 1965). The oxopolysaccharide (33) is treated with phenylhydrazine in dilute acetic acid solution. The insoluble coloured complex (34) formed contains approximately one molecule of phenylhydrazine condensed with each dialdehyde group. The complex breaks down when heated in an aqueous or ethanolic solution of phenylhydrazine and acetic acid, releasing phenyl-osazones (36) and the unoxidized parts of the molecule (35) which can be separated and identified. This treatment removes all the residues in the original molecule which contained α-glycol groups and new vicinal hydroxyl groups are exposed in the degraded polymer. The process can then be repeated, and in this way successive layers of residues are removed from the periphery of the molecule.

Some nitrogen remains in the degraded polysaccharide and the reducing end unit (35) may be present as a phenylosazone.

Successive applications of the degradation to the mucilage from *Cladophora rupestris* resulted in the isolation of glyoxal bisphenyl hydrazone after the first degradation indicating that 1,4-linked galactose and xylose units had been degraded, and after the third oxidation about 25% yield of a degraded polymer comprising galactose, arabinose and rhamnose residues was recovered (O'Donnell and Percival, 1959a). From this it can be concluded that

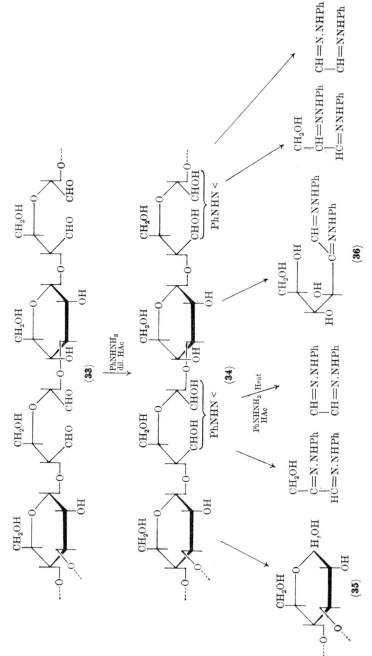

Fig. 7. Barry degradation.

the polysaccharide has a highly branched structure with xylose and galactose units at the ends of the branches and galactose, arabinose and rhamnose residues at the centre of the molecule. Since the recovered polymer still contained ester sulphate groups, it can be deduced that these groups are linked to residues both on the outer branches and in the centre of the molecule.

F. THE ACTION OF ALKALI AND DEDUCTIONS ON LINKAGE IN THE MOLECULE

Dilute alkali has a profound effect on reducing sugars causing degradation and complex dienol rearrangements yielding different saccharinic acids (Sowden, 1957). Extensive studies by Kenner, Richards and Corbett on the action of lime water in the absence of oxygen on partially methylated sugars and disaccharides revealed that the type of saccharinic acid resulting from this action depended on the point of attachment of the methyl group or the glycosidic link. For example, 3-O-methylglucose (37) gives a mixture of α- and β-*meta*saccharinic acids (41). Loss of proton results in the formation of an intermediate enolate ion (38). This is followed by β-elimination of the methoxide ion and the resulting enol (39) isomerises to the α-di-ketone (40) which then undergoes a benzilic rearrangement to form metasaccharinic acids (41). This reaction was used to confirm the 1,3-linkage in laminaran (Corbett and Kenner, 1955).

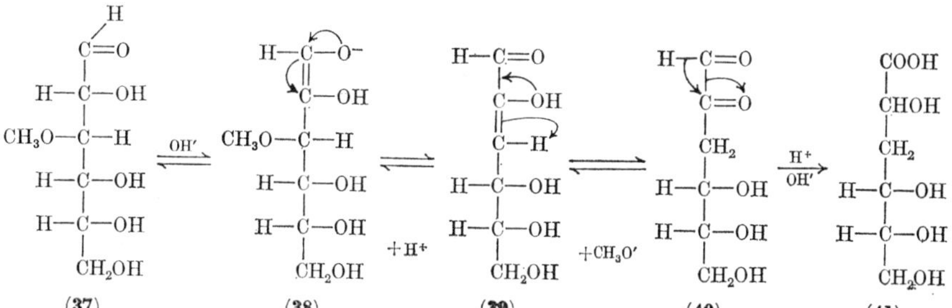

FIG. 8. The action of lime water on 3-O-Methylglucose.

G. THE USE OF ENZYMES IN STRUCTURAL STUDIES
(Manners, 1955, 1966;Whelan, 1961, 1966)

Although hydrolytic enzymes, polysaccharases, have played a valuable part in the elucidation of the structure of a number of algal polysaccharides, suitable degradative enzymes for many of these polymers have yet to be found. The enzymes use the polysaccharide as a substrate cleaving specific glycosidic linkages in the molecule; the extent of this reaction can be measured by (*a*) the decrease in viscosity, (*b*) the change in optical rotation, (*c*) the increase in reducing power and (*d*) chromatographic analysis of the products. The "action pattern" of the enzyme may be either by random cleavage of the chains (endo-splitting enzymes), or by stepwise hydrolysis starting at the non-reducing end of each of the polysaccharide chains (exo-splitting enzymes). In the latter case, every glycosidic link or alternate glycosidic link is broken with the production of mono- and di-saccharides, respectively. Random cleavage of glycosidic links leads to the formation of a series of oligosaccharides. If a branched polysaccharide is being investigated then random cleavage by a single polysaccharase leaves all the interchain linkages intact and a number of branched oligosaccharides of varying size results. A rapid fall in viscosity accompanied by a small increase in reducing power is indicative of random cleavage, while a slow decrease in viscosity with a rapid increase in reducing power and the early presence of mono- and di-saccharides is proof of endwise fission.

Highly purified, enzymically homogeneous preparations are essential for the structural analysis of polysaccharides. Other carbohydrases or transglycosylases are often associated with polysaccharases, but the advent of new techniques for the rigorous purification of proteins has removed any excuse for using other than pure enzymes in this type of work.

1. *Amylases*

Before use can be made of an enzyme in the determination of the fine structure of a polysaccharide, its mode of action must be determined by a study of the degradation of a polysaccharide of known structure. For example, the action pattern of salivary α-amylase and of the cereal β-amylase has been determined by using amylose and amylopectin as substrates. This knowledge has proved invaluable in the enzymic degradation of various algal starches (see pp. 75 to 82), and has enabled a direct comparison of these materials with land plant starches.

Salivary α-amylase catalyses the random hydrolysis of non-terminal α-1,4-linkages, the end products from amylose being maltose and maltotriose (Manners, 1955). The extent of the hydrolysis is determined by measuring the reducing power of the resulting solution, and is usually expressed as "apparent conversion into maltose" (P_M). α-Amylases cannot hydrolyse α-1,6-linkages and consequently the end-products from the action of salivary α-amylase on amylopectin are maltose, maltotriose, and branched α-dextrins (French, 1966). Again the extent of the degradation is determined by measuring the reducing power and expressing it as "apparent P_M". In contrast, β-amylolysis consists of stepwise hydrolysis of alternate linkages in a chain of α-1,4-linked glucose residues from the non-reducing ends, with the release of β-maltose. Enzyme action is arrested by the presence of branchpoints or of anomalous links in the chain (Manners, 1955, 1966). Linear amylose molecules are completely degraded by β-amylase, but amylopectin yields maltose and a β-limit dextrin of high molecular weight. Only the exterior chains are attacked, since β-amylase cannot by-pass the 1,6-linkages. It enables, therefore, amylose-type structures which comprise only α-1,4-glucosidic linkages to be distinguished from those containing other structural features such as α-1,3- and α-1,6-interchain linkages. Furthermore, the extent of β-amylolysis of amylopectin-type molecules is related to the average chain-length. Many amorphous β-amylase preparations contain a second amylolytic enzyme (Z-enzyme). It was originally suggested that Z-enzyme had a debranching action, but more recent studies have shown that Z-activity is caused by a trace of α-amylase inpurity and involves the random hydrolysis of one or two $(1 \rightarrow 4)$-α-D-glucosidic linkages, thus facilitating further β-amylolysis (Peat and Whelan, 1953).

Other enzymes have proved of value in structural analysis of starch. In particular the debranching enzymes, R-enzyme and *iso*amylase, have been used in the investigations on red and green algal amylopectins. These enzymes hydrolyse the 1,6-interchain linkages in amylopectin and the derived dextrin can then be further hydrolysed with β-amylase.

2. *Specific Algal Polysaccharases*

Enzymic hydrolysis is often more specific than acidic hydrolysis. Where several types of linkage are present, for example, 1,4- and 1,3-linkages in the galactans of the Rhodophyceae, certain bacterial enzymes will preferentially cleave the 1,4-links and leave the 1,3-links intact giving *neo*agarobiose (see p. 131) and *neo*carrabiose (see p. 139) from agar and carrageenan, respectively. Enzymes from a cell-free extract of a *Pseudomonad* randomly cleaved alginic acid yielding a series of oligosaccharides in which the non-reducing end unit derived from the point of cleavage is a 4,5-unsaturated uronic acid residue (see p. 112). Commercial hemicellulase randomly cleaved the mannan from *Codium fragile* (see p. 95) and enabled the separation of the β-1,4-linked homologous series of oligosaccharides. The structure of *Caulerpa* xylan as a linear β-1,3-linked xylan has been confirmed by its hydrolysis to xylose by the action of a β-1,3-xylanase (Fukui *et al.*, 1960).

H. THE SITE OF ESTER SULPHATE GROUPS

Many seaweed mucilages offer still more formidable difficulties owing to the presence of the half ester sulphate groups. They are very strong acids and attempted isolation or storage of the free acid polysaccharide invariably leads to complete degradation. Consequently such materials are generally isolated as their barium, sodium or ammonium salts.

1. Deductions from Acidic Hydrolysis

The sulphate group can be removed by mineral acid hydrolysis, but generally at the expense of cleavage of the glycosidic linkages (Turvey, 1965). Only in a few instances, for example, in the case of λ-carrageenan, porphyran and the mucilages from *Cladophora rupestris* and *Codium fragile*, has it been possible to isolate and characterize sulphated monosaccharides and so prove directly the site of the ester sulphate groups in the macromolecule.

Some algal polysaccharides contain both uronic acid and ester sulphate groups, and the presence of the former stabilizes the glycosidic linkages and enables the sulphate groups to be removed with dilute methanolic hydrogen chloride at room temperature with the minimum rupture of the glycosidic bonds, and the isolation of relatively large yields of degraded desulphated polysaccharides. By the application of periodate oxidation before and after sulphate removal, it has been possible in the mucilages from *Enteromorpha* and *Ulva* to show that the proportion of uncleaved rhamnose was greatly reduced in the oxopolysaccharide from desulphated material. This indicates that desulphation had produced additional glycol groups on the rhamnose units, and therefore that these units carried sulphate groups.

2. Deductions from Alkaline Hydrolysis

It has been established (Percival, 1949; Turvey, 1965) from experiments on model monosaccharide sulphates that the ester sulphate groups are

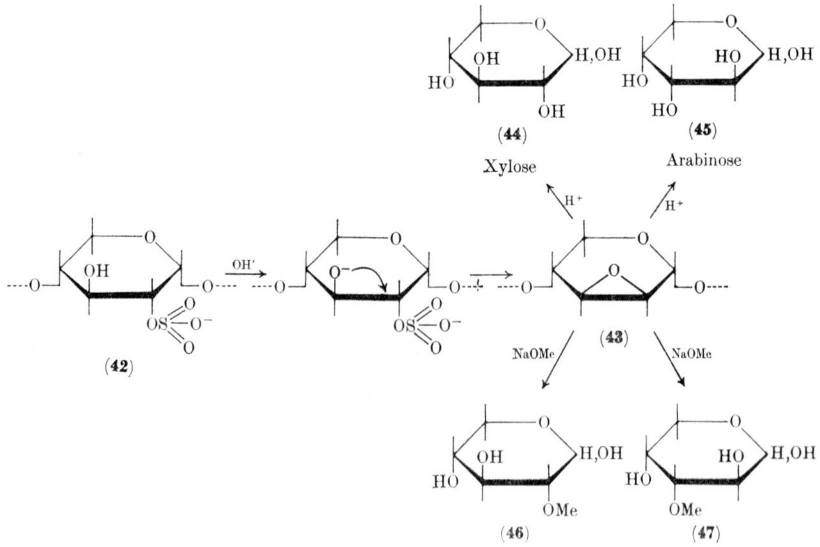

Fig. 9.

exceedingly stable to alkali unless (*a*) there is a free hydroxyl group adjacent and *trans* to the sulphate, as in the 2-sulphated xylose residues [Fig. 9 (**42**)] in the mucilage from *Ulva* (p. 184), when the sulphate is catalytically cleaved by the alkali with Walden Inversion and the formation of an epoxide ring (**43**), or (*b*) the sulphate is linked to C-3 or C-6 [Fig. 10, (**48**)], as in λ-carrageenan and porphyran, when cleavage occurs with the formation of 3,6-anhydro-galactose (**49**).

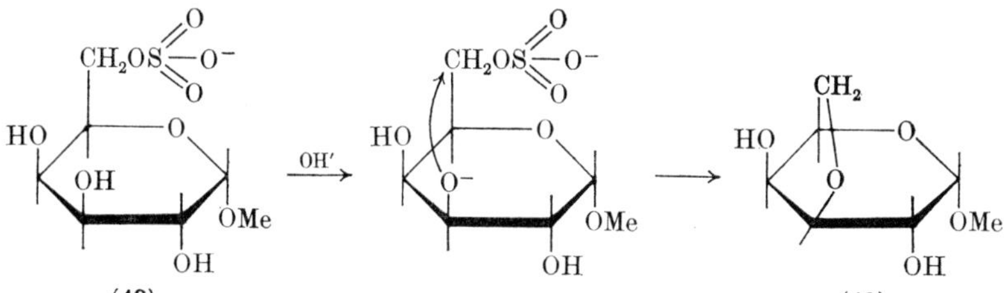

FIG. 10.

The epoxide rings on hydrolysis with acid or alkali may open on either side with the formation of the desulphated parent sugar (**44**) and a new sugar (**45**). Cleavage of the epoxide sugar with sodium methoxide likewise may give two monomethyl sugars (**46, 47**).

These facts have enabled deductions regarding the site of ester sulphate to be made where direct evidence was unobtainable. Hydrolysis of the alkali treated sulphated polysaccharide from *Ulva lactuca* revealed the presence of arabinose [Fig. 9 (**45**)], and a trace of lyxose, two sugars absent from the original polysaccharide (Percival and Wold, 1963). In a separate experiment, cleavage of the epoxide derivative with sodium methoxide gave 2-*O*-methyl-D-xylose [Fig. 9 (**46**)]. From these facts it was possible to deduce that the original polysaccharide contained some xylose 2-sulphate units. The formation of trace quantities of lyxose, which was identified as the D-sugar, is explained in terms of desulphation of D-xylose 2-sulphate terminal units [Fig. 11 (**50**)], and subsequent migration of the epoxide ring as shown in Fig. 11. The new ring would then undergo scission in the alkaline medium to give D-lyxose (**51**) and L-xylose (**52**).

3. *Diagnosis from Infrared Spectra*

Infrared studies have been helpful in the location of ester sulphate group-ings within some polysaccharides. Examination of model monosulphates of glucose and galactose have shown a broad absorption band at 1240 cm^{-1} characteristic of the S=O stretching vibration. In addition, specific bands

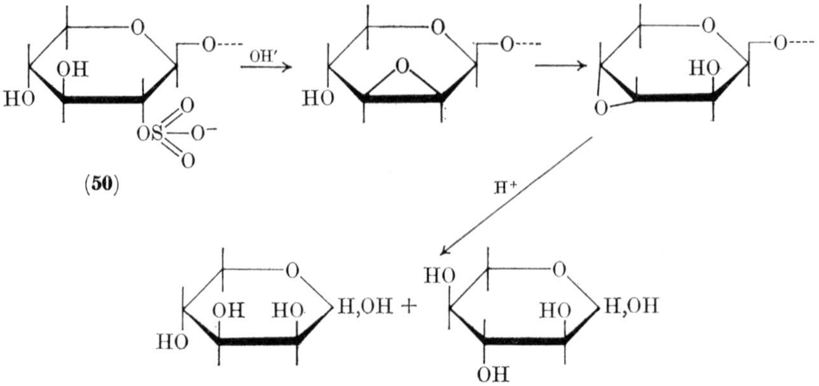

D-Lyxose
(51)

L-Xylose
(52)

FIG. 11.

corresponding to the C—O—S vibration, which are considered to be characteristic of the type of ester sulphate, for example, at 820 cm^{-1} (sulphated primary hydroxyl), at 830 cm^{-1} (sulphated equatorial hydroxyl) and at 850 cm^{-1} (sulphated axial hydroxyl) (Lloyd et al., 1961; Lloyd and Dodgson, 1961). The spectra of carrageenan and porphyran support the presence of more than one type of ester sulphate in the polysaccharides by showing two or more bands in the region 810—860 cm^{-1}. Bands at 850 cm^{-1} and 820 cm^{-1} shown by the sulphated polysaccharide from *Codium* (see p. 174) confirms the site of ester sulphate on C-4 and C-6 of the D-galactose units.

IV. THE DETERMINATION OF MOLECULAR WEIGHTS

Marine algal polysaccharides have molecular weights ranging from about 5000 to several millions. In general, a polysaccharide is made up of molecules of various sizes and in most cases average molecular weights have been reported. This can lead to some confusion in comparing results obtained by different methods, as the figure for the average molecular weight of a polydisperse polymer will depend on the method used (Greenwood, 1952; Whistler and BeMiller, 1965).

Some methods in effect count the number of molecules in a given weight of sample, and the figure obtained is termed the number average. In others, the effect of each molecule on the measured result is proportional to its weight, and a weight average is obtained. Unless all the molecules are of the same size, the weight average will always be higher than the number average, and some indication of the spread of molecular weights can be obtained by dividing the weight average by the number average. The greater the range of molecular sizes, the higher is this ratio.

To obtain detailed information on the distribution of molecular weights it is necessary to fractionate the polymer and to make determinations on the individual fractions. Few studies of this kind have been made on algal polysaccharides, but an example is the work on carrageenan fractions obtained by fractional precipitation with alcohol (Smith *et al.*, 1955) (see p. 146).

In considering the structure of a polysaccharide, the number of units in the molecule (the degree of polymerization or DP) is often more illuminating than the molecular weight; it is obtained by dividing the molecular weight of the polymer by that of the anhydro sugar unit.

A. CHEMICAL METHODS

The determination of reducing end-groups has been widely used as a means of molecular weight measurement in carbohydrate chemistry. In a structure built up of glycosidic links, there can be only one such residue per molecule.

Periodate oxidation can give information on both reducing and nonreducing end groups, and is therefore of particular value with branched chain molecules (see p. 36). These chemical methods give a number average molecular weight.

B. PHYSICAL METHODS

Osmotic pressure measurements and the associated method of isothermal distillation (Gee, 1940) give a direct value of the number average weight, and have been used with various algal polysaccharides (Donnan and Rose, 1950; Mackie and Percival, 1960).

On the other hand, light scattering determinations yield a weight average molecular weight and also give information on the dimensions of the polymer in solution (Stacey, 1956).

With the ultracentrifuge, various molecular weight averages may be obtained, depending on the method used. No simple average is obtained by rate of sedimentation and diffusion measurements, but it is possible to obtain a weight average during the approach to sedimentation equilibrium (Schachman, 1959).

The simplest and most widely used method of comparing the sizes of polymers, the measurement of viscosity, does not give a direct measurement of the molecular weight, but requires calibration by an absolute method for each polymeric type. Measurements can be expressed as the specific viscosity, i.e. the difference between the viscosity of the polymer solution and that of the solvent divided by the viscosity of the solvent, at different concentrations. The specific viscosity divided by the concentration is extrapolated to zero concentration, giving the ratio with the symbol $[\eta]$, generally referred to as the intrinsic viscosity. There is some ambiguity about its exact value, since while some workers have expressed the concentration as grams of polymer in 100 ml of solution others have used grams in 100 g of solution.[2]

The intrinsic viscosity of a polymer is a measure of the volume and shape of the space occupied by the molecule in solution; the molecular weight is only one factor in determining this, but the effect of other factors (such as the solvent used and the stiffness of the polymer chain) can be included with reasonable accuracy in a constant used with a polymer of one type in a specified solvent.

The Staudinger equation, which can be written as

$$[\eta] = KM$$

has been extensively used but has been found to require modification for many polymer solutions. The equation

$$[\eta] = KM^a$$

is more generally applicable, but the two constants K and a have both to be found by experiment.

Aqueous solutions of polyelectrolytes, common among algal polysaccharides, introduce further problems, and it is necessary to add a simple salt to the solution to obtain a linear extrapolation to zero concentration. The calculation of the molecular weight of alginic acid from sedimentation, diffusion and viscosity measurements (Cook and Smith, 1954) illustrates the uncertainties involved in these methods.

Viscosity measurements are of the greatest value in comparing different samples of the same polysaccharide, obtained from different sources by various methods.

REFERENCES

Aspinall, G. O. (1963). *J. chem. Soc.* 1676.
Barry, V. C. (1943). *Nature, Lond.* **152**, 537.
Binkley, W. W. (1955). *In* "Advances in Carbohydrate Chemistry" **10**, p. 55, Academic Press, New York and London.
Bishop, C. T. (1964). *In* "Advances in Carbohydrate Chemistry" **19**, p. 95, Academic Press, New York and London.

[2] The "limiting viscosity number" also with the symbol $[\eta]$ was introduced by a IUPAC committee in 1952 (*J. Polymer Sci.* **8**, 257). Here the concentration is given in grams/ml of solution so that the limiting viscosity number is approximately 100 times the intrinsic viscosity. It has not yet been widely used in publications on polysaccharides.

Bouveng, H. O., and Lindberg, B. (1960). *In* "Advances in Carbohydrate Chemistry" **15**, p. 53, Academic Press, New York and London.

Brown, E. G., and Hayes, T. J. (1952). *Analyst, Lond.* **77**, 445.

Brown, H. C., and Subba Rao, B. C. (1960). *J. Am. chem. Soc.* **82**, 681.

Cantley, M., Hough, L., and Pittet, A. O. (1959). *Chemy. Ind.* 1126.

Cook, W. H., and Smith, D. B. (1954). *Can. J. Biochem. Physiol.* **32**, 227.

Corbett, W. M., and Kenner, J. (1955). *J. chem. Soc.* 1431.

Donnan, F. G., and Rose, R. S. (1950). *Can. J. Res.* B**28**, 105.

Fisher, I. S., and Percival, Elizabeth (1957). *J. chem. Soc.* 2666.

French, D. (1966). *Biochem. J.* **100**, 2P.

Fukui, S., Suzuki, T., Kitahara, K., and Miwa, T. (1960). *J. gen. appl. Microbiol., Tokyo* **6**, 270.

Gee, G. (1940). *Trans. Faraday Soc.* **36**, 1164.

Goldstein, I. J., Hay, G. W., Lewis, B. A., and Smith, F. (1965). *In* "Methods in Carbohydrate Chemistry" (R. L. Whistler and J. N. BeMiller, eds), Vol. 5, p. 361, Academic Press, New York and London.

Greenwood, C. T. (1952). *In* "Advances in Carbohydrate Chemistry" **7**, p. 289, Academic Press, New York and London.

Gregory, J. D. (1960). *Archs Biochem. Biophys.* **89**, 157.

Haq, Q. N., and Percival, Elizabeth (1966). *In* "Some Contemporary Studies in Marine Science" (H. Barnes, ed), p. 355, Allen and Unwin, London.

Haug, A., and Smidsrod, O. (1965). *Acta chem. scand.* **19**, 1221.

Hay, G. W., Lewis, B. A., and Smith, F. (1965a). *In* "Methods in Carbohydrate Chemistry" (R. L. Whistler and J. N. BeMiller, eds), Vol. 5, p. 357, Academic Press, New York and London.

Hay, G. W., Lewis, B. A., and Smith, F. (1965b) *In* "Methods in Carbohydrate Chemistry" (R. L. Whistler and J. N. BeMiller, eds), Vol. 5, p. 377, Academic Press, New York, and London.

Hay, G. W., Lewis, B. A., Smith, F., and Unrau, A. M. (1965c). *In* "Methods in Carbohydrate Chemistry" (R. L. Whistler and J. N. BeMiller, eds), Vol. 5, p. 251, Academic Press, New York and London.

Hirst, E. L., and Percival, Elizabeth (1965). *In* "Methods in Carbohydrate Chemistry" (R. L. Whistler and J. N. BeMiller, eds), Vol. 5, p. 287, Academic Press, New York and London.

Hirst, E. L., Percival, Elizabeth, and Wold, J. K. (1964). *J. chem. Soc.* p. 1493.

Hirst, Sir Edmund, Mackie, W. and Percival, Elizabeth, (1965). *J. chem. Soc.* 2958.

Hjerten, S. (1962). *Biochim. biophys. Acta* **62**, 445.

Jones, J. K. N. (1950). *J. chem. Soc.* 3292.

Jones, J. K. N., and Perry, M. B. (1957). *J. Am. chem. Soc.* **79**, 2787.

Jones, J. K. N., and Pridham, J. B. (1953). *Nature, Lond.* **172**, 161.

Kaye, M. A. G., and Kent, P. W. (1953). *J. chem. Soc.* 79.

Kowkabany, G. N. (1954). *In* "Advances in Carbohydrate Chemistry" **9**, p. 304, Academic Press, New York and London.

Larsen, B., Haug, A., and Painter, T. J. (1966). *Acta chem. scand.* **20**, 219.

Lewis, B. A., and Smith, F. (1963). *Abstr. Pap. Am. chem. Soc.* **144**, 8D.

Lloyd, A. G., Dodgson, K. S., Price, R. G., and Rose, F. A. (1961). *Biochim. biophys. Acta* **46**, 108.

Lloyd, A. G., and Dodgson, K. S. (1961). *Biochim. biophys. Acta* **46**, 116.

Love, J., and Percival, Elizabeth (1964a). *J. chem. Soc.* 3338.

Love, J., and Percival, Elizabeth (1964b). *J. chem. Soc.* 3345.

McDowell, R. H. (1958). *Chemy. Ind.* 1401.

Mackie, I. M., and Percival, Elizabeth (1960). *J. chem. Soc.* 2381.

Mackie, I. M., and Percival, Elizabeth (1961). *J. chem. Soc.* 3010.

Manners, D. J. (1955). *Q. Rev. chem. Soc.* **9**, 73.

Manners, D. J. (1966). *Biochem. J.* **100**, 2P.

Meier, H. (1958). *Acta chem. scand.* **12**, 144.

Nevell, T. P. (1963). *In* "Methods in Carbohydrate Chemistry" (R. L. Whistler and J. N. BeMiller, eds), Vol. 3, p. 39, Academic Press, New York and London.

O'Colla, P. S. (1965). *In* "Methods in Carbohydrate Chemistry" (R. L. Whistler and J. N. BeMiller, eds), Vol. 5, p. 382, Academic Press, New York and London.

O'Donnell, J. J., and Percival, Elizabeth (1959a). *J. chem. Soc.* 1739.

O'Donnell, J. J., and Percival, Elizabeth (1959b). *J. chem. Soc.* 2168.

Peat, S., and Whelan, W. J. (1953). *Nature, Lond.* **172**, 492.

Percival, E. G. V. (1949). *Q. Rev. chem. Soc.* **3**, 369.

Percival, Elizabeth, and Wold, J. K. (1963). *J. chem. Soc.* 5459.

Rees, D. A., and Samuel, J. W. B. (1965). *Chemy. Ind.* 200.

Schachman, H. K. (1959). "Ultracentrifugation in Biochemistry", Academic Press, New York and London.

Schoch, T. J. (1945). *In* "Advances in Carbohydrate Chemistry" **1**, p. 247, Academic Press, New York and London.

Scott, J. E. (1960). *In* "Methods of Biochemical Analysis" (D. Glick, ed.), Vol. 8, p. 145, Interscience Publishers, New York.

Smith, D. B., and Cook, W. H. (1953). *Archs. Biochem. Biophys.* **45**, 232.

Smith, D. B., O'Neill, A. N., and Perlin, A. S. (1955). *Can. J. chem.* **33**, 1352 and ref. cited therein.

Sowden, J. C. (1957). *In* "Advances in Carbohydrate Chemistry" **12**, p. 35, Academic Press, New York and London.

Stacey, K. A. (1956). "Light Scattering in Physical Chemistry", Butterworths, London.

Sweeley, C. C., Bentley, R., Makita, M., and Wells, W. W. (1963). *J. Am. chem. Soc.* **85**, 2497.

Turvey, J. R. (1965). *In* "Advances in Carbohydrate Chemistry" **20**, p. 183, Academic Press, New York and London.

Weigel, H. (1963). *In* "Advances in Carbohydrate Chemistry" **18**, p. 61, Academic Press, New York and London.

Whelan, W. J. (1961). *Nature, Lond.* **190**, 954.

Whelan, W. J. (1966). *Biochem. J.* **100**, 1P.

Whistler, R. L., and BeMiller, J. N. (1965) "Methods in Carbohydrate Chemistry" Vol. 5, Section V, Academic Press, New York and London.

Yamakawa, T., and Ueta, N. (1964). *Jap. J. exp. Med.* (Engineering) **34**, 37.

CHAPTER 3

Food Storage Polysaccharides of the Phaeophyceae and Crysophyceae

Laminaran, earlier named laminarin, a water-soluble β-glucan, is the food reserve material of the Phaeophyceae and also of the Chrysophyceae. The Rhodophyceae, however, appear to be devoid of this type of polysaccharide, the main food reserve being floridean starch, an α-glucan (see Chapter 4). Although tentative evidence for the presence of a small amount of a laminaran-type polysaccharide has been reported for *Caulerpa filiformis* (Mackie and Percival, 1959) (see p. 92) and for *Cladophora rupestris* (see p. 165) (Fisher and Percival, 1957), both members of the Chlorophyceae, the main storage material of this class of algae is a starch-type polysaccharide which closely resembles that in land plants (see Chapter 4).

I. LAMINARAN FROM THE PHAEOPHYCEAE

Laminaran, abundant in a wide variety of brown algae (Quillet, 1958), was first isolated by Schmiedelberg in 1885. It is absent from the stipe of the *Laminarias*, from the frond during periods of rapid growth in the spring, and and from the actively growing regions adjacent to the stipe, but increases in quantity up the frond (Black, 1954). In the autumn and winter between 20 and 36% of the dry weight of the frond can be laminaran (Black and Dewar, 1949). Seasonal variations are considerably less in the *Fucales*, and are most marked in *Fucus serratus* which grows near the low tide level. In this plant it varies from 2 to 10%, but in other *Fucales*, growing higher up the beach, variations are less regular and the amounts found do not exceed about 7% (Black, 1948, 1949). Recent extensive studies of the laminaran content of sixteen genera of brown algae found on the Pacific coast of N. America (Powell and Meeuse, 1964) have shown that the best sources there are *Laminaria saccharina* and *Alaria* with laminaran 22 to 34% of the dry weight.

A. ISOLATION

Laminaran occurs in two forms distinguished by their solubility in cold water. They are therefore referred to as "soluble" and "insoluble" laminaran, although both forms dissolve in hot water.

3+

This polysaccharide is particularly susceptible to attack by moulds and bacteria (Dillon, 1964), and hence extraction of seaweed which has been stored in conditions of moderate moisture should be avoided.

1. Cold-water Insoluble Form (Black, 1965)

Minced or finely shredded fronds of *Laminaria hyperborea* are extracted with 0·09N-hydrochloric acid for 30 min at 70°, the residual weed is filtered through muslin and washed with warm water. The combined filtrate and washings are stirred for 3 hr and set aside for 3 days. The snowy white precipitated laminaran is collected by centrifugation and washed with alcohol and ether. If dried milled weed is used, the extraction can be carried out at room temperature in the presence of a small volume (1·5 ml/litre) of 40% formaldehyde. Yield about 3% of wet weight. After neutralization and concentration of the centrifugate, crude soluble laminaran may be obtained by treatment of the resulting solution with ethanol to 85% concentration. This can be purified from contaminating fucoidan by the method given below for the soluble laminaran from *L. digitata*.

The insoluble laminaran can be recrystallized from a 10% aqueous solution in 82% recovery, $[\alpha]_D$ –12 to –14°.

2. The Cold-water Soluble Form

This is usually extracted from dried, milled *L. digitata* frond (laminaran content about 24% dry weight) in a similar manner. After stirring in 0·09N-hydrochloric acid in the cold for 2 hr, the solution is centrifuged, and the weed residues washed with 0·05N-hydrochloric acid. The combined filtrate and washings are made up to 85% concentration with ethanol. The precipitate is washed with ethanol and ether and air-dried. The derived crude laminaran (about 70% of the total) is freed from contaminating fucoidan by dissolution in water (15·5 g in 180 ml) and passage of the solution through a Zeo-Karb 225 resin column (20 ml). The column is washed with water (50 ml) and the combined effluent is treated with ethanol to 85% concentration. The laminaran separates as a semicolloid and is coagulated by the addition of sodium chloride (0·2 g in 1 ml water). The white precipitate ($[\alpha]_D$–12° in water) is collected by centrifugation and dried under reduced pressure, and has ash, 1·3%, laminaran, 82·2%, and moisture about 10%.

B. ANALYSIS AND QUANTITATIVE DETERMINATION

Hydrolysis with dilute mineral acid (N-sulphuric acid for 8 hr at 100°) (Barry, 1938) gives glucose as the sole reducing carbohydrate in a number of samples (Beattie *et al.*, 1961), but small proportions of mannose have also been reported (Smith and Unrau, 1959; Chesters and Bull, 1963b). However, examination of a range of samples revealed a maximum mannose content of only 0·2%. This amount is structurally insignificant and is probably formed by epimerization during extraction and purification.

Determination of the laminaran content (Cameron et al., 1948) of a particular weed is based on the reducing power of such a hydrolysate. Since other polysaccharides such as alginic acid and fucoidan also produce reducing sugars, it is necessary to measure the reducing power before and after removal of the glucose. This may be done by the micro-method of Shaffer and Somogyi on as little as one gram of ground seaweed of known moisture content. The glucose can be removed in about eight minutes with an excess of washed yeast cells (Harding and Selby, 1931). The writer suggests as an alternative method for the determination of glucose the use of glucose oxidase (Salton, 1960).

Laminaran is also susceptible to hydrolysis by snail enzymes and enzymic hydrolysis can be used in quantitative determinations, the specificity and completeness of the reaction being carefully controlled by paper chromatography (Quillet, 1958).

C. DETERMINATION OF THE FINE STRUCTURE

Early methylation studies (Barry, 1939), together with the low negative rotation, indicated that laminaran is an essentially linear β-1, 3-linked glucan, since 2,4,6-tri-O-methyl-D-glucose was the major methylated sugar obtained on hydrolysis of the methylated material. This was confirmed by partial acidic and enzymic hydrolysis which gave a homologous series of β-1,3-linked

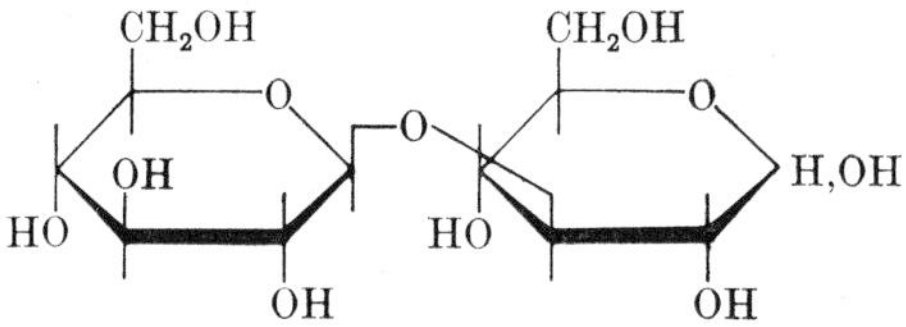

(1)
Laminaribiose

oligosaccharides from which the disaccharide, laminaribiose (1) was separated and characterized (Bächli and Percival, 1952). Further confirmation that at least part of the molecule consisted of 1,3-linked units was obtained by the isolation of a 40–50% yield of D-glucometasaccharinic acid [Fig. 1 (3)] (a product diagnostic of 1,3-linkage) (see p. 43), after treatment of laminaran (2) with lime water (Corbett and Kenner, 1955). Theory for a linear β-1,3-linked glucose polysaccharide requires 100% metasaccharinic acid (Fig. 1, but see page 57).

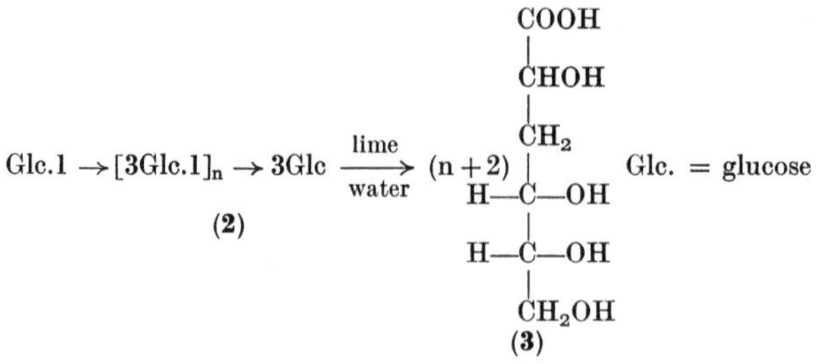
Glc.1 → [3Glc.1]n → 3Glc
lime
water
(n + 2)
COOH
CHOH
CH2
H—C—OH
H—C—OH
CH2OH
(2)
(3)
Glc. = glucose

FIG. 1. The action of lime water on a linear chain of 1,3-linked glucose units.

1. *The Presence and Linkage of Mannitol in Laminaran*

An important advance in our understanding of the laminaran molecule was the discovery by the Bangor School of 1·7% and 2·7% of mannitol as a constituent in insoluble (*L. hyperborea*) and soluble (*L. digitata*) laminaran, respectively (Peat *et al.*, 1958). After partial acid hydrolysis of insoluble laminaran, these authors separated small quantities of mannitol, 1-*O*-β-D-glucosylmannitol, 1-*O*-laminaribiosylmannitol and gentiobiose (**23**) in addition to much larger amounts of glucose, laminaribiose and higher laminarisaccharides. From the proportion of mannitol it was concluded that about 40% of the molecules are terminated by a mannitol residue (M-chains) [Fig. 2 (**4**)] linked through one of the two primary alcohol groups, (due to the

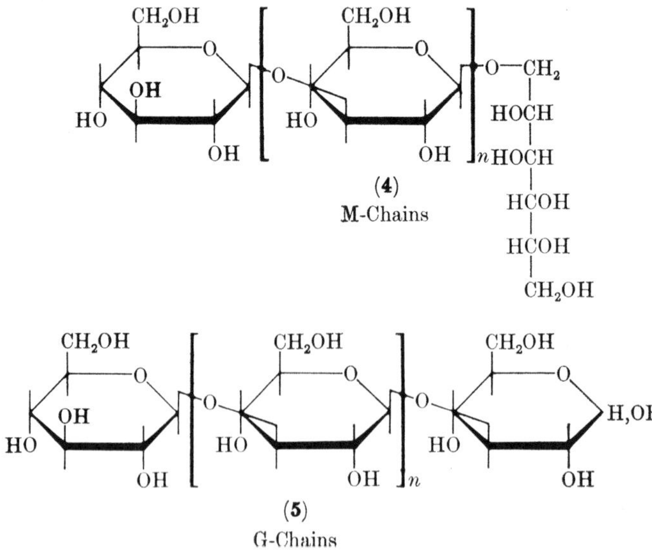
CH2OH
OH
HO
OH
O—CH2
HOCH
HOCH
HCOH
HCOH
CH2OH
n
(4)
M-Chains
H,OH
(5)
G-Chains

FIG. 2. Types of molecule found in laminaran.

symmetry of the mannitol molecule, C-1 and C-6 are the same with respect to the rest of the molecule), and the rest (G-chains) [Fig. 2 (5)] are terminated by reducing glucose residues linked through C-3.

Later work (Anderson *et al.*, 1957) indicated 75% M-chains in a sample of the soluble form and 46% in the insoluble material but see also Fleming *et al.* (1966).

The M-chains would be immune to alkaline degradation and to "over-oxidation" by periodate (see p. 36) and their presence in laminaran explains why only 0·5 mole of saccharinic acid per residue was obtained on alkaline degradation, and why the "overoxidation" by periodate was arrested when about 50% of the material had been degraded (Hough and Perry, 1956).

On oxidation of about 0·4% solution of laminaran in 15 mM-sodium meta-periodate at 2°, the M-chains rapidly release 1 mole of formaldehyde and this renders measurement of the mannitol content comparatively easy, since under these conditions of oxidation, the G-chains do not yield formaldehyde (see p. 39). Oxidation of a number of samples of laminaran from different species of *Laminaria* showed a range of 2 to 3% in the mannitol content (Anderson *et al.*, 1958; Annan *et al.*, 1965a). The accuracy of the method was confirmed by hydrolysis of two samples and separation of the glucose and mannitol.

The inability to detect ethylene glycol after periodate oxidation, reduction and hydrolysis [Fig. 3, (6–8)] of laminaran (M-chains) led to the suggestion

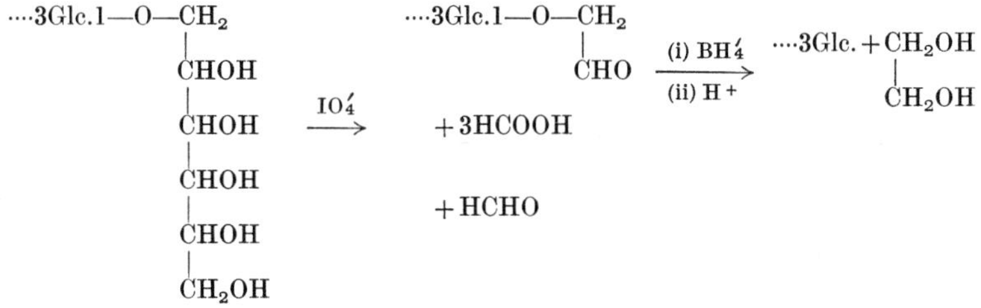

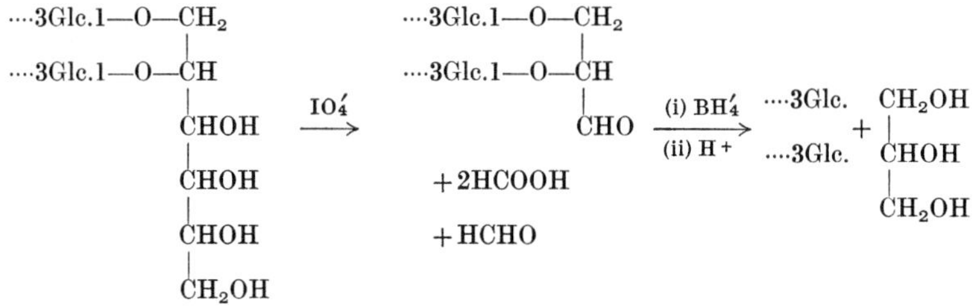

FIG. 3.

by Goldstein *et al.*, (1959) that the mannitol residues are disubstituted through C-1 and C-2 [Fig. 3. (**9**)]. Such residues would yield glycerol (**11**) under this treatment. Improved chromatographic techniques, however, revealed, on repetition of this experiment, the presence of ethylene glycol (**8**), and after separation on resin and cellulose it was isolated and characterized (Annan *et al.*, 1962).

Further confirmation that mannitol is singly linked through C-1 (or C-6) was obtained by determination of the formic acid released on oxidation of about 0·025% solution of laminaran in 0·4 mM-sodium metaperiodate at 2° (Clancy and Whelan, 1959). Under these exceptionally mild conditions, G-chains do not yield formic acid; M-chains singly linked through C-1 (or C-6) should produce three moles of formic acid [Fig. 3 (**7**)] and chains with mannitol linked through C-1 and C-2 (C-6 and C-5) (**9**) would give only two moles of formic acid per mannitol residue (**10**). It was found (Annan *et al.*, 1965a) that three different samples of laminaran gave 3·0 to 3·1 moles of formic acid per mannitol residue.

Additional confirmation was obtained by gas-liquid chromatographic detection of 2,3,4,5,6-penta-*O*-methylmannitol in a methanolysate of methylated laminaran.

Treatment of laminaran with borohydride converts the reducing glucose units (G-chains) into D-glucitol (sorbitol) (Abdel-Akher *et al.*, 1951; Fleming and Manners, 1965), and the product is known as laminaritol (Fig. 4.). It comprises mannitol terminated chains (M-chains) linked to C-1 of the mannitol residue and sorbitol terminated chains (S-chains) linked to C-3 of the sorbitol.

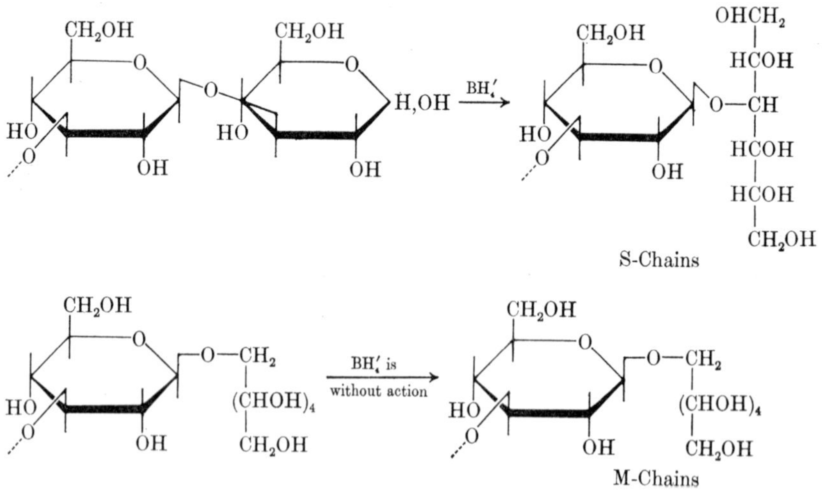

Fig. 4. Laminaritol.

Similarly, oxidation of laminaran with bromine converts the reducing glucose units into gluconic acid (laminaric acid) (Fig. 5) without affecting the M-chains, and it is then possible to separate the two molecular species on an anion-exchange resin (Goldstein *et al.*, 1959). The unoxidized M-chains (designated by these authors laminaritol, which is not to be confused with the more general use of this term for reduced whole laminaran—see above), are not absorbed on the resin whereas, the oxidized G-chains (termed laminaric acid) are.

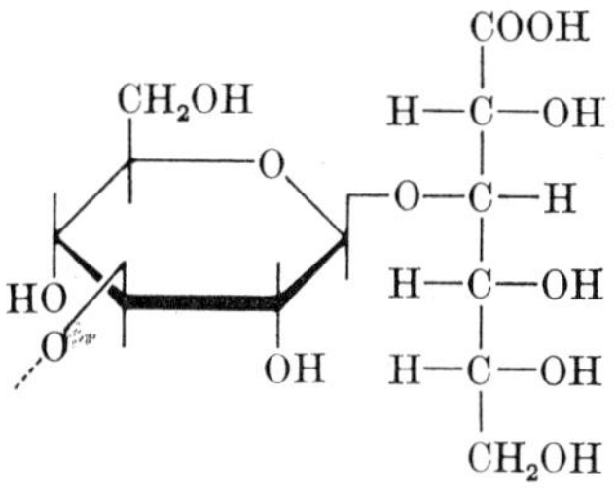

FIG. 5. Laminaric acid.

2. *The Position of 1,6-Glucosidic Linkages*

The small quantity (0·26%) of gentiobiose (**23**) together with 6-*O*-β-laminaribiosylglucose and 3-*O*-β-gentiobiosylglucose obtained on partial hydrolysis of insoluble laminaran indicated that laminaran contained 1,6-glucosidic linkages, but the proportion of these was very small in the sample examined since the molar ratio of laminaribiose to gentiobiose was 70:1. Unfortunately, these results do not tell us whether the 1,6-linkages occur as inter-residue linkages or as branch points, and evidence in favour of both locations has been advanced.

In favour of 1,6-inter-residue linkage is:

(*a*) 3:6-Di-*O*-β-D-glucosyl-D-glucose is absent in the hydrolysis products of laminaran, but it should be pointed out that hydrolyses of amylopectin or glycogen have failed to yield any significant quantities of the analogous 4,6-di-*O*-α-D-glucosyl-D-glucose (Peat *et al.*, 1955; Thompson and Wolfrom, 1951).

(*b*) The recovery of ethylene glycol after subjection of laminaran to Smith degradation (see p. 39) followed by a second reduction and complete hydrolysis was considered to favour inter-residue 1,6-linkages [Fig. 6 (**12–16**)] (Smith and Unrau, 1959), but did not exclude the possibility of 1,6-branch points as well.

But, as previously shown, the ethylene glycol could have been derived from the singly linked mannitol (**20–22**).

(*c*) Further support for 1,6-inter-residue linkages was advanced from the

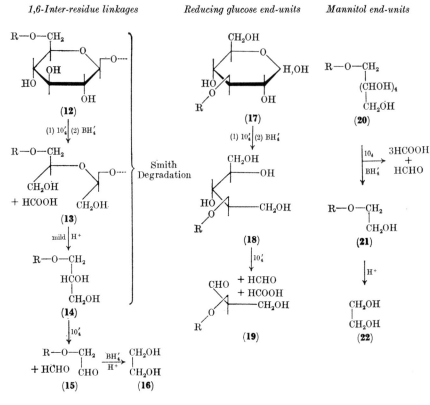

1,6-Inter-residue linkages   *Reducing glucose end-units*   *Mannitol end-units*

FIG. 6.

chromatographic identification of 2,3,4-tri-O-methylglucose in the hydrolysis products of methylated laminaran (Unrau, 1959; Beattie *et al.*, 1961), but again it is possible that this could have been derived from the tetramethylglucose by demethylation during hydrolysis (Annan *et al.*, 1965b).

In contrast, it is argued that 1,6-linkages are present as branch points since:

(*a*) Repeated application of Barry degradation (see p. 41) (Hirst *et al.*, 1958) failed to produce fragments of the polysaccharide small enough to pass through a cellophane membrane. This indicates that if any 1,6-inter-residue linkages are present, they must be near the ends of the chains. The polysaccharide would be cleaved at any 1,6-inter-residue by this treatment.

(*b*) While the presence of 1,6-inter-residue linkages would also cause fragmentation of the molecule when subjected to Smith degradation [Fig. 6. (**12–14**)], application of this series of reactions to insoluble laminaran from *L. hyperborea* or to soluble laminaran from *L. digitata* caused no fragmentation

(Annan *et al.*, 1965b; Fleming *et al.*, 1966) so that 1,6-linkages in the main chains appear to be absent from these samples.

(*c*) The Smith degradation sequence of reactions converts reducing glucose end groups into 2 substituted D-arabitol residues [Fig. 6 (**18**)], at the same time 1,6-linked glucose residues (**12**) are cleaved with the production of 3-substituted glycerol residues (**14**). A second periodate oxidation of these products should, therefore, liberate one molecule of formaldehyde from each original reducing glucose residue (**19**) and another molecule from every 1,6-inter-residue linkage (**15**). M-Chains do not yield formaldehyde on the second oxidation [the formaldehyde released from the first oxidation having been lost in the isolation of the polyalcohol (**21**) before mild hydrolysis and the second oxidation].

In practice, it was found that a sample of laminaran of DP 24 which contained 43% of G-chains gave 0·19 mole of formaldehyde per anhydrohexose residue. Theory for this sample in the absence of 1,6-inter-residue linkages requires 0·18 molecule per anhydrohexose residue. This result, therefore, supports other evidence for the presence of 1,6-branch points and the absence of a significant proportion of 1,6-inter-residue linkages. It appears to the writers that the overall evidence on samples of insoluble laminaran from *L. hyperborea* and soluble laminaran from *L. digitata* favours 1,6-branch points rather than 1,6-linkages in the main chain.

3. *Attempts to Distinguish Soluble from Insoluble Laminaran*

In the early chemical investigations on laminaran (Connell *et al.*, 1950), parallel experiments were carried out on the insoluble and soluble forms in an attempt to discover any structural differences between them. Acetylation, followed by methylation of purified samples of the two polysaccharides, gave results which were indistinguishable. Separation of the methylated hydrolysates gave about 5% of tetra-*O*-methylglucose (later studies, Anderson *et al.*, 1958, gave 4·4%) which corresponds to an average chain length of 20 to 23 glucose units. Nevertheless, the reducing power of soluble laminaran corresponded to about 57 units whereas that of insoluble laminaran was equivalent to about 40 units. Periodate oxidation released a slightly lower proportion of formic acid and higher yield of formaldehyde from the insoluble form (Percival and Ross, 1951), indicating a higher proportion of M- chains in the latter (contrast p. 56). However, later work (Fleming *et al.*, 1966) shows that there is little consistency in the proportion of M-chains in the two forms.

More recent investigations were carried out to determine whether there is any difference in the degree of branching in the two forms. This was done by comparing the average chain length (CL) with the degree of polymerization (DP) (Annan *et al.*, 1965b; Fleming *et al.*, 1966).

In the absence of 1,6-inter-residue linkages, the amount of formic acid

3*

released on mild periodate oxidation (one mole from each non-reducing end group, see Fig. 7) is a measure of the average CL if allowance is made for the formic acid arising from the mannitol residues. This gave average CL values for samples of soluble (*L. saccharina*) and insoluble (*L. hyperborea*) laminaran of 11 and 19, respectively.

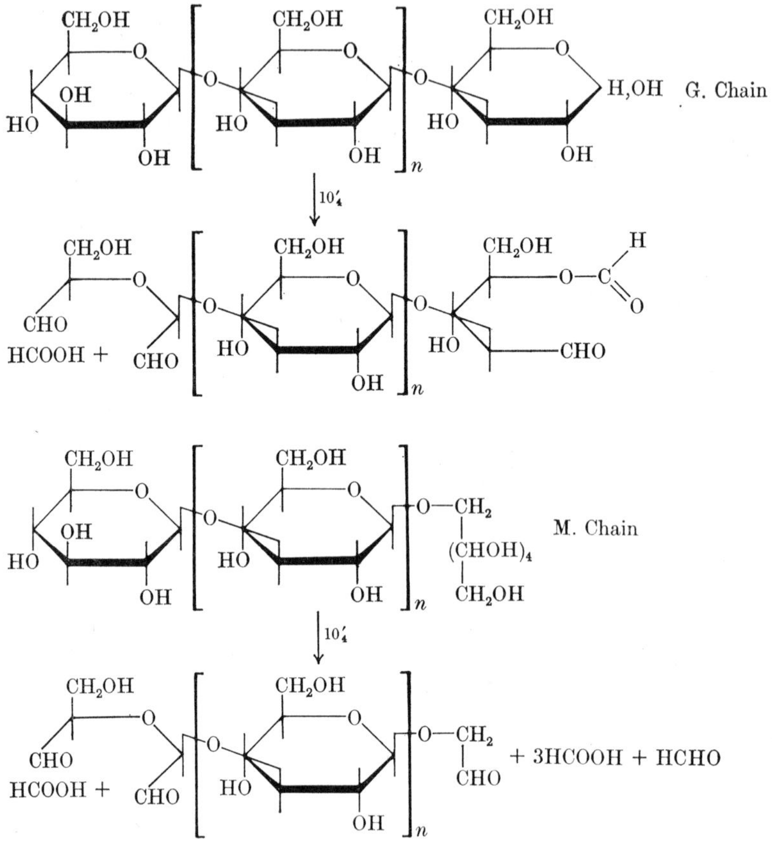

FIG. 7. Formic acid released on mild periodate oxidation of Laminaran.

It had been found that the average DP of laminaran can not be calculated from the amount of formaldehyde released on oxidation with periodate. Even with the less soluble potassium salt or with a limited excess of sodium metaperiodate at room temperature, oxidation does not stop with the liberation of one mole of formaldehyde from each type of chain. Complete oxidation of the G-chains occurs in a stepwise manner (see p. 38) with the release of a mole of formaldehyde from each residue.

Experiments, therefore, were carried out on reduced laminaran (laminaritol, see Fig. 4) in which the G-chains had been converted into S-chains. It was expected that the derived 3-linked sorbitol residues in laminaritol would release two moles of formaldehyde on periodate oxidation (Unrau and Smith, 1957). However, investigation of 3-O-methyl-D-glucitol (sorbitol) (Cantley et al., 1963) revealed that oxidation at pH 3·6 (the pH normally used) gave only 1·5 moles of formaldehyde. Experiments with model compounds (Clancy and Whelan, 1959) showed that it was possible to release one mole of formaldehyde only per mole of sugar alcohol if the reaction was carried out at room temperature with very dilute periodate. Similarly, laminaribitol released one mole of formaldehyde (determined by extending the reaction for 7 hr and extrapolating back to zero time) when oxidized with 1·43mM-periodate at room temperature (Annan et al., 1965b). Applying these latter conditions to samples of laminaritol prepared from the above soluble (L. saccharina) and insoluble (L. hyperborea) laminaran, the release of formaldehyde 0·036 and 0·042 moles per anhydro unit corresponded to average DP values of 28 and 24, respectively.

These values for CL and DP indicate a statistical average of 0·3 and 1·6 branches per molecule for these particular samples of insoluble and soluble laminaran, respectively. Similar studies on different samples of insoluble laminaran, four from L. hyperborea and one commercial product, and on samples of soluble laminaran, two from L. digitata and one from Fucus serratus (Fleming and Manners, 1965), gave average CL and DP values for the insoluble laminarans which were virtually identical. The CL values for the soluble polysaccharides, however, ranged from 7 to 10 and the DP values from 26 to 31 indicating an average of 2 to 3 branch points per molecule. These results are in agreement with the differences in reducing power and release of formic acid by these two forms of laminaran. The higher degree of branching in the soluble form explains its greater solubility since the presence of branches hinders close packing, and hence hydrogen bonding, between the individual molecules.

It should be emphasized that all the structural investigations described so far have been on laminaran isolated from species of Laminaria and Fucus. Study of the laminarans from other genera, although it will probably reveal the same overall basic structure of mainly β-1,3-linked glucose units, may show considerable variations in the fine details of structure. An excellent example of this is found in the laminaran isolated from the brown seaweed, Eisenia bicyclis. It has $[\alpha]_D$–45·5°, $[\eta]$ 1·35, is very soluble in cold water, and is devoid of mannitol. This absence of mannitol was confirmed by the fact that the DP (18) found by measuring the reducing power was essentially the same as that (21) determined cryoscopically. Proof also for the presence of a considerable proportion of 1,6-inter-residue linkages was obtained (Handa and

Nisizawa, 1961) both from methylation and partial hydrolysis studies. Hydrolysis of the methylated material gave 2,3,4,6-tetra-*O*-methyl, 2,3,4-tri-*O*- and 2,4,6-tri-*O*-methyl-glucoses as the sole products indicating an essentially linear structure of 1,3- and 1,6-linked glucose units. Partial hydrolysis led to the separation and identification of gentiobiose (**23**), laminaribiose (**1**), gentiotriose, 6-β-laminaribiosylglucose, laminaritriose and 3-β-gentiobiosylglucose together with tentative identification of some tetrasaccharides including gentiotetraose. Periodate oxidation studies confirmed these findings.

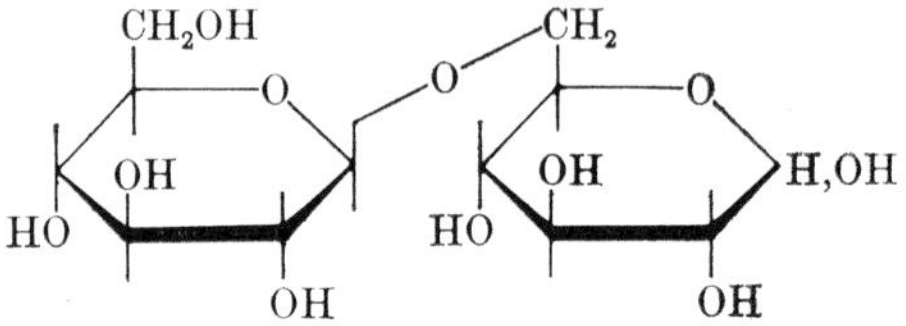

(**23**)

Gentiobiose

Alkaline degradation ceased after the loss of about one-third of the material (cf. *Laminaria* laminaran, p. 57), but in this case, it is the presence of a 1,6-linkage in the chain that arrests the degradation.

On the overall evidence, the Japanese authors suggest that *Eisenia* laminaran is a linear β-glucan with a DP of about 20 containing both 1,3- and 1,6-inter-residue linkages in the approximate proportion of 2:1.

D. MOLECULAR WEIGHT DETERMINATIONS

Comparisons of solutions of methylated laminaran and octaacetyl sucrose in chloroform by a modification of Barger's method (Caesar *et al.*, 1947) gave molecular weights between 2600 and 3500 for insoluble laminaran (*L. hyperborea*) (Connell *et al.*, 1950) and 2800 and 3800 for soluble laminaran (*L. digitata*) (Percival and Ross, 1951). A provisional estimate by osmotic pressure on the insoluble form gave a value between 3000 and 5000, although there was some evidence of diffusibility through the membrane.

Molecular weights, calculated from sedimentation, diffusion and viscosity data, of soluble laminaran from *L. digitata* and insoluble laminaran from *L. hyperborea*, were approximately 5300 and 3500, respectively, and it was shown that both materials were polydisperse (Friedlaender *et al.*, 1954). These authors also found that partly methylated insoluble laminaran (OMe, 2·7 and 6·9%) had molecular weights of 3700 and 2900, respectively, indicating degradation during methylation. Fractional precipitation of methylated laminaran (Broatch and Greenwood, 1956) led to the separation of material which by isothermal distillation (Gee, 1940) was shown to have a number average molecular weight of 12,000 (average DP 58). Examination of laminaran and lime-treated laminaran in the Spinco ultracentrifuge using the synthetic boundary cell technique gave Schlieren patterns which suggested that the laminaran had a very wide range of molecular weights. Lime-treated laminaran, on the other hand, was more homogeneous and possessed a larger sedimentation constant, 1.0×10^{-13} c.g.s. units, compared with 0.5×10^{-13} c.g.s. units for laminaran.

E. Enzymic Hydrolysis of Laminaran

The enzymes which hydrolyse laminaran are termed laminarases (= laminarinases) and Bull and Chesters (1966) have recently prepared an excellent review on the biochemistry of laminaran and the nature of laminarase in which they suggest that this term should be used to describe the whole enzyme complex which degrades laminaran, including exo- and endo-hydrolytic β-1, 3-glucanases and β-D-glucosidases.

1. Occurrence of Laminarase

Laminarase occurs in such widely diverse organisms as bacteria, fungi, algae, higher plants, and molluscs. These enzymes are involved in the intra-cellular mobilization of food reserves (not only in algae, but also in higher plants and fungi all of which synthesize β-1,3-glucans) and are also encountered in the extracellular breakdown of plant debris and in the digestive metabolism of invertebrates.

Enzyme preparations have been reported from cereals, hyacinth bulbs and potato tubers (Dillon and O'Colla, 1950), from brown, red, and green seaweeds (Duncan et al., 1956); from Euglena (Fellig, 1960), from marine bacteria, from many fungi (Chesters and Bull, 1963a; 1963b) and from the digestive juices of molluscs (Barry, 1941; Nisizawa, 1939; Quillet, 1958).

2. Isolation and Method of Assay

A. Enzymes produced extracellularly by bacteria and fungi are readily isolated from cold cell-free filtrates at pH 5-6 by precipitation with acetone, ethanol, or ammonium sulphate, although Chesters and Bull (1963c) found that the concentrations of ammonium sulphate necessary for precipitation caused severe inactivation of the fungal enzyme, and, therefore, used only acetone precipitated preparations. The resultant creamy-white, freeze-dried material can be stored at low temperature and retains its activity for considerable periods. Amylase activity may be removed from fungal preparations by adsorption on carboxymethylcellulose (Chesters and Bull, 1963b), and protamine sulphate precipitation has been used to remove nucleic acids from other laminarase preparations (Tanaka, 1964).

B. Freshly collected seaweed is minced in ice-water and extracted with dilute sodium carbonate solution at room temperature, the extract dialysed and the enzymes precipitated with ammonium sulphate (Duncan et al., 1956). After dissolution in water and reprecipitation, the enzymes are dissolved in water, the solution dialysed until free from ammonium sulphate and freeze-dried.

C. Homogenized plant and fungal tissues are most effectively extracted with buffer solution, and from molluscs the enzyme may be extracted from the digestive gland juices or from homogenized gut and gut content.

The purest laminarase preparations have been obtained from Bacillus circulans (Horikoshi et al., 1963) and Sclerotina libertiana (Ebata and Satormura, 1963). Resins were used to purify both these preparations and that from Sclerotina was crystallized from calcium acetate solution by additions of cold acetone; it was specific for β-1,3-glucosidic links and attacked only laminaran and sclerotan.

The majority of these preparations consist of a mixture of endo- and exo-β-1,3-glucanases (for explanation of endo and exo see p. 44) and β-glucosidases. Calcium phosphate has been used to resolve laminarase from Streptomyces species S 93 (Lester, 1958), which comprised 50 and 31%, respectively, of two endo-β-glucanases and 19% of an exo-β-glucanase. Fractionation of several fungal laminarases has also been achieved on cellulose ion exchangers and purification on hydroxyapatite. Low endo- and exo-hydrolytic activity was found in extracts of a number of brown, red and green marine algae (Duncan et al., 1956).

The laminarase activity is generally assayed by measuring the reducing sugar (as glucose) produced from laminaran under standard conditions and the activity defined as 1U produces reducing sugar equivalent to 1 mg of glucose from a 0·5% solution of laminaran at pH 5·8 and 37° in 30 minutes (Chester and Bull, 1963a), under which conditions the reaction kinetics are first order. An alternative viscometric assay method using carboxymethylpachyman has also been reported (Clarke and Stone, 1962).

3. Specificity and Action Pattern

In general, laminarases are multienzyme systems. Analysis of 128 fungal enzymes (Chesters and Bull, 1963b) revealed that 25% contained exo-, 56% contained endo- and 19% contained mixed-hydrolytic activity. For example, Penicillum stipitatum H127 laminarase consisted only of an endo-hydrolytic enzyme whereas that from Myrothecium verrucaria comprised three endo- and one exo-hydrolytic enzymes. In addition one or more β-glucosidases are often present.

Extracts from wheat, oats, barley, potato and hyacinth bulbs (Dillon and O'Colla, 1951) apparently have an endo-hydrolytic effect on laminaran, yielding oligosaccharides, while the various seaweed extracts which contain a number of other enzymes in addition to exo- and endo-β-1,3-glucanases (Duncan et al., 1956) produce from laminaran glucose in addition to a series of oligosaccharides. The two components in the extract from Cladophora rupestris are differentiated by the greater heat lability of the endo-β-1,3-glucanase.

Several different enzymes take part also in the breakdown of other β-glucans. For example, the degradation of barley β-glucan involves at least three enzymes: endo- and exo-β-glucanases and a β-glucosidase (Bass et al., 1953; Preece and Hoggan, 1956).

All the laminarases exhibit complete specificity for β-1,3-glucans, but the β-glucosidase components determine the range of components attacked. The specificity is influenced by the conformation of the glucosidic linkages, the carbon atoms involved and the chain length of the substrate. It has been shown that laminarase preferentially hydrolyses 3-substituted β-glucosyl units rather than β-1,3-linkages, and this may explain the reported inactivity of a β-1,3-glucanase from Bacillus subtilis to laminaran (Moscatelli et al., 1961).

Both soluble and insoluble laminaran give the same hydrolysis pattern. Endwise attack (exo-enzyme) consists in the removal of glucose units from the non-reducing ends of the chains (Reese and Mandels, 1959), whereas attack by endo-β-glucanases consists of random cleavage and the release of a series of laminaridextrins (Peat et al., 1952; Dillon and O'Colla, 1950, 1951; Manners, 1955). Although exo-hydrolysis produces glucose as the sole initial product, prolonged incubation yields laminaridextrins as a result of chain shortening. Mannitol-containing oligosaccharides were detected with the Trichoderma and Myrothecium enzymes (Chesters and Bull, 1963b) and laminaribiosylmannitol and β-glucosylmannitol were very resistant to further hydrolysis. 3-O-β-Gentiobiosyl-D-glucose showed transitory accumulation and tended to disappear, while gentiobiose accumulated, but this may be due in part to trans β-glucosylase activity.

Enzymic activity increases with laminaran concentrations up to 5 to 6 mg/ml and extrapolation of published data gave K_m values in the range 0·5 (Lester, 1958) to 3·5 mg (Luchsinger et al., 1963) similar to those of β-1,2-, β-1,4-, and β-1,6-glucanases (Reese and Mandels, 1963).

4. *Optimal Conditions for Activity and Inhibitory Factors*

Fungal laminarase activity occurred over the temperature range 10 to 60° (Chesters and Bull, 1963c) with an optimum at 37°. Examination of the separated components of a Streptomyces preparation revealed an optimal temperature for exo-β-1,3-glucanase of 27 to 30° and for endo-1,3-β-glucanase

37 to 42°, and this higher temperature for the latter type of fungal glucanase appeared to be universal. However, it should be pointed out that the temperature ranges are to some extent dependent upon pH conditions (Chesters and Bull, 1963c; Reese and Mandels, 1959). Exo-β-1,3-glucanases from fungae have pH optima ranging from 4·9 to 5·0 and the corresponding endo-enzymes from pH 6·0 to 6·1. The laminarase activities of extracts of *C. rupestris*, *Rhodymenia palmata* and *U. lactuca* are optimum at about pH 5·5, 6·0 and 6·3, respectively. Some twenty fungal laminarases showed three types of pH activity over the pH range 3·2 to 7·9. Some gave a single peak of activity over this pH range, while others gave two roughly equal peaks and again others gave a large and a small peak. This is to be expected when it is remembered that many of these are mixtures of enzymes, each peak representing the optimum of a particular enzyme (Chesters and Bull, 1963b). This may be considered as further proof of the multienzyme nature of laminarases.

Little is known of the thermostability of laminarases (Bull and Chesters, 1966).

Both exo- and endo-β-1,3-glucanases are stimulated by Fe^{3+}, Mn^{2+} and Co^{2+} ions at 1mM concentration, and the enzyme from *Euglena* is activated by 50 to 100% by Mn^{2+} and to some extent by Co^{2+}, although Fe^{3+} appears to be a mild inhibitor (Fellig, 1960). Such activation may be due to a loose binding of the metal to the enzyme and substrate, whereby the configuration or energy of the molecule is altered, and so the rate of the reaction increased. This idea is supported by exhaustive dialysis, and by the fact that chelating agents such as EDTA have little effect on the enzyme. Malt laminarase activity is increased by nearly a third by sodium chloride and to a lesser extent by potassium chloride, sodium sulphate and dibasic sodium phosphate (Ferrell and Luchsinger, 1964; Luchsinger *et al.*, 1963). It seems that either an ionic environment favours enzyme action or a changing surface potential of the substrate leads to increased activity.

Heavy metals, such as Cu^{2+}, Ag^+ and Hg^{2+} and phenyl mercurinitrate, were highly inhibitory to the fungal laminarases (Chesters and Bull, 1963c) although some doubt has been cast on the inhibitory effect of the last substance (Clarke and Stone, 1965). Reese and Mandels (1960) and Clarke and Stone (1965) in comprehensive surveys showed that glucono $(1 \rightarrow 5)$ lactone was a very effective inhibitor of β-glucosidases but was, on the whole, without action on β-glucanases, and in the few cases where it appeared to be inhibitory it was considered that this was due to the effect on one only of the mixture of enzymes. This conclusion was supported by the results of Chesters and Bull (1963c) who found that the inhibitory effect of $(1 \rightarrow 4)$ gluconolactone on fungal laminarases was dependent upon the amount of exo-β-1,3-glucanase in the mixture.

F. PROPERTIES, DERIVATIVES AND USES

Laminaran is a white, odourless, tasteless powder which is insoluble in all common organic solvents. Although the insoluble form is insoluble in cold water, it is soluble in the hot (22 g in 100 g water at 90°). Both forms give neutral, non-viscous solutions which do not form gels, and do not possess adhesive or film-forming properties. The rate constant of hydrolysis of insoluble laminaran has been determined at three different temperatures and three different concentrations of hydrochloric acid (Szejtli, 1965). The entropy of activation was found to be 9·19 cal/mol.

Water insoluble and water soluble forms yield the same derivatives. The hydroxyethyl, hydroxypropyl and benzyl ethers have been prepared in good yield, also acetate, benzoate and carbanilate esters, the latter in quantitative yield (Black and Dewar, 1954). β-Amino-ethyl ether derivatives containing 0·5 to 1·0 residues per monosaccharide unit have been prepared employing ethylenimine (O'Neill, 1955).

By treating laminaran with a solution of chlorosulphonic acid in liquid sulphur dioxide at −20°, preparations, showing no noticeable degradation and containing 1·7 sulphate groups per glucose unit, were obtained. Similar sulphation of the β-aminoethyl derivatives gave products containing both sulphamic and half ester sulphate groups.

Polymers with a molecular weight 4000 to 8000 have been shown to be most suitable as blood anticoagulants, and laminaran has therefore been considered as a possible material. It was necessary first to sulphate it since a second essential criterion for anticoagulant activity is the presence of ester sulphate groups. The neutral sodium salts of sulphated laminaran and its β-amino-ethyl derivative (see above) were tested for their anticoagulant activities in comparison with a standard heparin. The nitrogen derivative containing 46·2% sulphate was found to have the highest activity, 40 IU/mg compared with 100 for heparin.

In vivo studies have shown that laminaran sulphate has about one third the activity of heparin when injected into dogs (Hawkins and Leonard, 1958; Dewar, 1956). It is absorbed like heparin from the intestine of the rat to only a very small extent, and also like heparin, is excreted in the faeces after a large oral dose. Laminaran sulphate reduces the turbidity of alimentary lipaemia in rats and dogs, and it has been shown by Besterman and Evans (1957) to act similarly in humans toward hyperlipaemia in ischaemic heart disease.

II. LAMINARAN FROM THE BACILLARIOPHYCEAE

Only recently, as a result of improved culture techniques and the development of better chemical methods, has it been possible to investigate the polysaccharides synthesized by these minute organisms. *Phaeodactylum tricornutum*, a microscopic green marine diatom belonging to the Bacillario-physeae class of algae and the Phylum Chrysophyta, was earlier confused with *Nitzschia closterium*. It constitutes a food for zooplankton and is an import-ant constituent in the diets of mussels and oysters. When grown under bac-teria-free conditions with atmospheric carbon dioxide as the sole carbon source, it was found to synthesize a water-soluble glucan of the laminaran type which comprised about 14% of the dry weight of the organism.

The glucan was extracted with cold water and precipitated from low mole-cular weight material with ethanol (Ford and Percival, 1965). Hydrolysis showed contamination with trace quantities of xylose, mannose, and rham-nose and purification was achieved by fractionation on a column of DEAE-cellulose. Paper chromatography of a partial acid hydrolysate revealed spots with the mobilities of glucose, laminaribiose, gentiobiose and laminaritriose, and treatment with an endo-β-1,3-glucanase gave a similar chromatographic pattern. Parallel partial acidic and enzymic hydrolyses on soluble *Laminaria* laminaran established the essential similarity of the two materials, the major difference being the absence of mannitol in the *Phaeodactylum* polysaccharide.

Methylation confirmed these findings. The major methylated sugar, 2,4,6-tri-*O*-methylglucose was separated as a crystalline material and small quan-tities of 2,3,4,6-tetra-*O*- and 2,4-di-*O*-methylglucoses were identified by paper and gas-liquid chromatography and by ionophoresis.

P. tricornutum glucan reduced under the same conditions a higher propor-tion (0·53 mole) of periodate than *Laminaria* laminaran (0·39 mole). This could be due to the fact that "overoxidation" occurs in all the chains whereas the M-chains in laminaran resist "overoxidation".

The fact that glycerol was the sole cleaved product of Smith degradation (see p. 39) is further evidence that *Phaeodactylum* laminaran comprises a β-1,3-linked glucan with some branching at C-6, and, as is to be expected, closely resembles chrysolaminaran (Beattie *et al.*, 1961) the food reserve polysaccharide of fresh-water diatoms.

It seems that marine algal laminaran is not a single molecular species, but that the name covers a whole range of essentially linear β-1,3-linked glucans, in some of which the reducing end is terminated by a mannitol residue while others have reducing glucose units. Laminaran from some species is a mixture of these two types of molecule while that from other species is completely devoid of mannitol. The presence of a high proportion of 1,6-inter-residue linkages has been proved for the laminaran from *Eisenia bicyclis*, and a small proportion of these linkages may also be present in the laminaran from *Laminaria* and *Fucus*, spp. Most samples appear to have a small degree of branching at C-6. The presence of C-6 inter-residue linkages and of branch points seems to determine the solubility of the polysaccharide in cold water.

Where the proportion of these is very small, then the laminaran precipitates out of cold water.

Sufficient data on a wide variety of laminarans from different genera and species are not yet available to conclude whether or not these variations in fine structure are bound up with the particular environment of the plant.

REFERENCES

Abdel-Akher, M., Hamilton, J. K., and Smith, F. (1951). *J. Am. chem. Soc.* **73**, 4691.
Anderson, F. B., Hirst, E. L., and Manners, D. J. (1957). *Chemy. Ind.* p. 1178.
Anderson, F. B., Hirst, E. L., Manners, D. J., and Ross, A. G. (1958). *J. chem. Soc.* p. 3233.
Annan, W. D., Hirst, E. L., and Manners, D. J. (1962). *Chemy. Ind.* p. 984.
Annan, W. D., Hirst, Sir E. L., and Manners, D. J. (1965a). *J. chem. Soc.* p. 220.
Annan, W. D., Hirst, Sir E. L., and Manners, D. J. (1965b). *J. chem. Soc.* p. 885.
Bächli, P., and Percival, E. G. V. (1952). *J. chem. Soc.* p. 1243.
Barry, V. C. (1938). *Sci. Proc. R. Dubl. Soc.* **21**, 615.
Barry, V. C. (1939). *Sci. Proc. R. Dubl. Soc.* **22**, 59.
Barry, V. C. (1941). *Sci. Proc. R. Dubl. Soc.* **22**, 423.
Bass, E. J., Meredith, W. O. S., and Anderson, J. A. (1953). *Cereal Chem.* **30**, 313.
Beattie, A., Hirst, E. L., and Percival, Elizabeth (1961). *Biochem. J.* **79**, 531.
Besterman, E. M. M., and Evans, J. (1957). *Brit. Med. J.* No. **5014**, 310.
Black, W. A. P. (1948). *J. Soc. chem. Ind., Lond.* **67**, 355.
Black, W. A. P. (1949). *J. Soc. chem. Ind., Lond.* **68**, 183.
Black, W. A. P. (1954). *J. mar. biol. Ass. U.K.* **33**, 49.
Black, W. A. P. (1965). *In* "Methods in Carbohydrate Chemistry" (R. L. Whistler and J. M. BeMiller, eds), Vol. 5, p. 159, Academic Press, New York and London.
Black, W. A. P., and Dewar, E. T. (1949). *J. mar. biol. Ass. U.K.* **28**, 673.
Black, W. A. P., and Dewar, E. T. (1954). *J. Sci. Fd. Agric.* **5**, 176.
Broatch, W. N., and Greenwood, C. T. (1956). *Chemy. Ind.* p. 1015.
Bull, A. T., and Chesters, C. G. C. (1966). *In* "Adv. in Enzymology" (F. F. Nord, ed.), Vol. 28, p. 325, Interscience Publishers, N.Y., London and Sydney.
Caesar, G. V., Gruenhut, N. S., and Cushing, M. L. (1947). *J. Am. chem. Soc.* **69**, 617.
Cameron, M. C., Ross, A. G., and Percival, E. G. V. (1948). *J. Soc. Chem. Ind.* **67**, 161.
Cantley, M., Hough L., and Pittet, A. O. (1963). *J. chem. Soc.* p. 2527.
Chesters, C. G. C., and Bull, A. T. (1963a). *Biochem. J.* **86**, 28.
Chesters, C. G. C., and Bull, A. T. (1963b). *Biochem. J.* **86**, 31.
Chesters, C. G. C., and Bull, A. T. (1963c). *Biochem. J.* **86**, 38.
Clancy, M. J., and Whelan, W. J. (1959). *Chemy. Ind.* p. 673.
Clarke, A. E., and Stone, B. A. (1962). *Phytochem.* **1**, 175.
Clarke, A. E. and Stone, B. A. (1965). *Biochem. J.* **96**, 793.
Connell, J. J., Hirst, E. L., and Percival, E. G. V. (1950). *J. chem. Soc.* p. 3494.
Corbett, W. M. and Kenner, J., (1955). *J. chem. Soc.* p. 1431.
Dewar, E. T. (1956). *Proc. 2nd int. Seaweed Symp.*, Trondheim, 1955 (T. Braarud and N. A. Sörensen, eds), p. 55, Pergamon Press, Lond., N.Y., Paris.
Dillon, T. (1964). *Proc. 4th int. Seaweed Symp.*, Biarritz, 1961 (A. D. deVirville and J. Feldmann, eds), p. 301, Pergamon Press, Lond., N.Y.
Dillon, T., and O'Colla, P. (1950). *Nature, Lond.* **166**, 67.
Dillon, T., and O'Colla, P. (1951). *Chemy. Ind.* p. 111.
Duncan, W. A. M., Manners, D. J., and Ross, A. G. (1956). *Biochem. J.* **63**, 44.

Ebata, J., and Satormura, Y. (1963). *Agric. biol. Chem.* (Tokyo) **27**, 471.

Fellig, J. (1960). *Science, N.Y.* **131**, 832.

Ferrell, W. J., and Luchsinger, W. W. (1964). *Proc. W. Va Acad. Sci.* **36**, 1.

Fisher, I. S., and Percival, Elizabeth (1957). *J. chem. Soc.* p. 2666.

Fleming, M., and Manners, D. J. (1965). *Biochem. J.* **94**, 17P.

Fleming, M., Hirst, E. L., and Manners, D. J. (1966). *Proc. 5th int. Seaweed Symp.*, Halifax, Nova Scotia (E. G. Young and J. L. McLachlan, eds), p. 255, Pergamon Press, Oxford.

Ford, C. W., and Percival, Elizabeth (1965). *J. chem. Soc.*, p. 7035.

Friedlaender, M. G. H., Cook, W. H., and Martin, W. G. (1954). *Biochem. biophys. Acta* **14**, 136.

Gee, G. (1940). *Trans. Faraday Soc.* **36**, 1164.

Goldstein, I. J., Smith, F., and Unrau, A. M. (1959). *Chemy. Ind.* p. 124.

Handa, N., and Nisizawa, K. (1961). *Nature, Lond.* **192**, 1078.

Harding, V. J., and Selby, D. L. (1931). *Biochem. J.* **25**, 1815 and ref. cited therein.

Hawkins, W. W., and Leonard, V. G. (1958). *Can. J. Biochem. Physiol.* **36**, 161.

Hirst, E. L., O'Donnell, J. J., and Percival, Elizabeth (1958). *Chemy. Ind.* p. 834.

Horikoshi, K., Koffler, H., and Arima, K. (1963). *Biochim. biophys. Acta* **73**, 267.

Hough, L., and Perry, M. B. (1956). *Chemy. Ind.* 768.

Lester, R. (1958). Doctoral Thesis, University of Nottingham.

Luchsinger, W. W., Ferrell, W. J., and Schneberger, G. L. (1963). *Cereal Chem.* **40**, 554.

Manners, D. J. (1955). *Proc. Biochem. Soc.* **61**, p. xiii.

Mackie, I. M., and Percival, Elizabeth (1959). *J. chem. Soc.* p. 1151.

Moscatelli, E. A., Ham, E. A., and Rickes, E. L. (1961). *J. biol. Chem.* **236**, 2858.

Nisizawa, K. (1939). *J. chem. Soc.*, Japan **60**, p. 1020.

O'Neill, A. N. (1955). *Can. J. Chem.* **33**, 1097.

Peat, S., Thomas, G. J., and Whelan, W. J. (1952). *J. chem. Soc.* p. 722.

Peat, S., Whelan, W. J., and Edwards, T. E. (1955). *J. chem. Soc.* p. 355.

Peat, S., Whelan, W. J., and Lawley, H. G. (1958). *J. chem. Soc.* p. 724; p. 729.

Percival, E. G. V., and Ross, A. G. (1951). *J. chem. Soc.* p. 720.

Powell, J. H., and Meeuse, B. D. J. (1964). *Econ. Bot.* **18**, 164.

Preece, I. A., and Hoggan, A. (1956). *J. Inst. Brew.* **62**, 485

Quillet, M. (1958). *C. r. hebd. Séanc. Acad. Sci.*, Paris. **246**, 812.

Reese, E. T., and Mandels, M. (1959). *Can. J. Microbiol.* **5**, 173.

Reese, E. T., and Mandels, M. (1960). *Develop Ind. Microbiol*, **1**, 171.

Reese, E. T., and Mandels, M. (1963). *In* "Advances in Enzymic Hydrolysis, Cellulose and Related Materials" (E. T. Reese, ed.), p. 197, Pergamon Press, New York.

Salton, M. R. J. (1960). *Nature, Lond.* **186**, 966 and ref. cited therein.

Smith, F., and Unrau, A. M. (1959a). *Chemy. Ind.* p. 636.

Smith, F., and Unrau, A. M. (1959b). *Chemy. Ind.* p. 881.

Szejtli, J. (1965). *Acta chim. hung.* **45**, 141.

Tanaka, H. (1964). Doctoral thesis, University of California, Davis, California, U.S.A.

Thompson, A., and Wolfrom, M. L. (1951). *J. Am. chem. Soc.* **73**, 5849.

Unrau, A. M. (1959). Ph. D. thesis, University of Minnesota, U.S.A.

Unrau, A. M., and Smith, F. (1957). *Chemy. Ind.* p. 330.

CHAPTER 4

Other Neutral Polysaccharides, Food Reserve and Structural

Food Reserve

I. ALGAL STARCHES

A. FLORIDEAN STARCH

The food reserve of the Rhodophyceae is in the form of characteristically shaped granules which stain red-violet with iodine, a colour similar to that given by animal glycogens. Kylin (1913) examined the grains in forty different species of algae and first isolated the grains from *Furcellaria fastigiata*, and showed that they comprised a glucan (now known as floridean starch) structurally related to starch since they could be hydrolysed by malt diastase.

Extensive structural studies have been carried out on the floridean starch from *Dilsea edulis*. The starch-type polysaccharide can be extracted from the weed with hot water, but at the same time galactan sulphates (see p. 127) are also dissolved out. In an attempt to overcome this difficulty, O'Colla (1953) removed much of the sulphated mucilage with cold acid before extraction with hot water, and purification of the starch was achieved by resin treatment. The resulting material, however, still gave 18% of galactose on hydrolysis. Nevertheless, after methylation and hydrolysis a 42% yield of 2,3,6-tri-O-methyl-D-glucose was separated, and 48% of maltose was obtained after incubation of the extract with wheat β-amylase, indicating a high proportion of α-1,4-glucosidic linkages. Attempts to separate an amylose fraction by butanol, amyl alcohol and thymol were unsuccessful.

Examination of a sample of *Dilsea edulis* starch which was essentially free from contaminating galactan (Fleming *et al.*, 1956) gave on hydrolysis 93% of reducing sugar and had the properties detailed in Table 1 (B). It reduced 1·05 mole of periodate per glucose residue and the amount of formic acid liberated by potassium metaperiodate indicated an average chain length of nine glucose residues. These results support the earlier work and provide evidence for the presence of α-1,6-branch points in the macromolecule and confirm the amylopectin-glycogen nature of this polysaccharide.

Peat and his co-workers (1959) showed that floridean starch [Table 1 (A)] from *D. edulis* was indeed closely related to the amylopectins of land plants. By treating one gram portions of the extract by the iodine precipitation method (Steiner and Guthrie, 1944), these authors succeeded in freeing the extract from contaminating galactan. The proportion of the starch not pre-

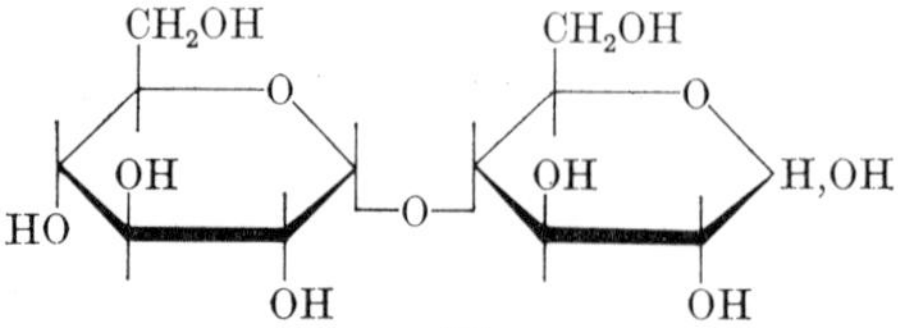

(1)
Maltose

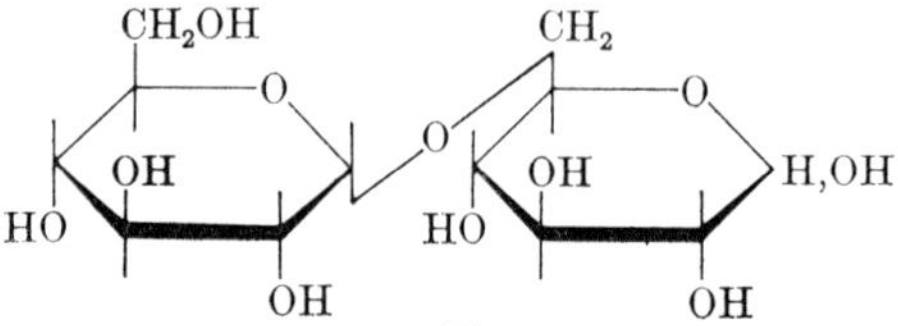

(2)
*Iso*maltose

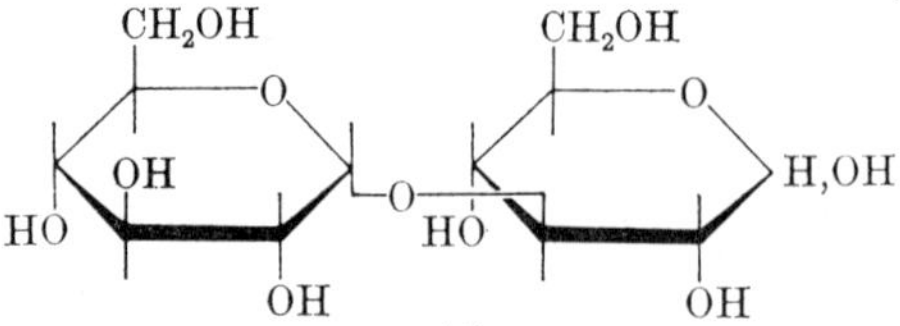

(4)
Nigerose

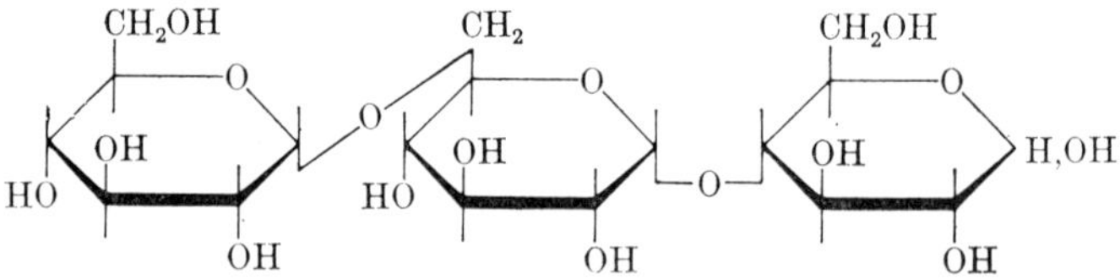

(3)
Panose
(6-*O*-α-D-Glucopyranosylmaltose)

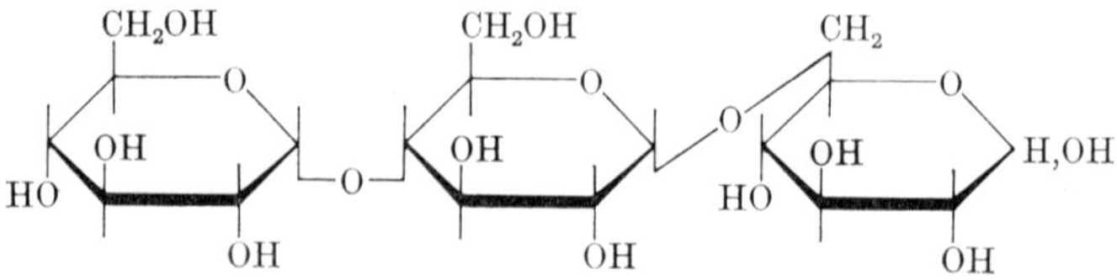

(5)
6-*O*-α-Maltosylglucose
(4-*O*-α-D-Glucopyranosyl*iso*maltose)

FIG. 1.

cipitated as the iodine complex was recovered with the galactan by ethanol precipitation after the excess iodine had been removed by sodium thiosulphate. This was then separated from the galactan by precipitation of the latter as the cetyltrimethylammonium bromide complex (see p. 28). The recovered starch after removal of inorganic material by dialysis contained 92·6% glucose, 0·25% ash and 0·24% nitrogen. On partial acid hydrolysis, it yielded glucose and maltose (1) with lesser amounts of *iso*maltose (2), panose (3), nigerose (4) and 6-*O*-α-maltosylglucose (5). All the disaccharides were characterized as their crystalline acetates.

The resemblance of this sample of floridean starch to amylopectin was supported by the action of β-amylase, α-amylase, R-enzyme and by comparison of its properties with those of glycogen and waxy maize starch (See Table 1). The presence of nigerose (4) (34·6 mg from 12·5 g starch) in the acid hydrolysate in greater quantity than might be expected from reversion synthesis, suggested the presence of a small proportion of α-1,3-links, in addition to the α-1,4- and α-1,6-linkages. This was confirmed by the enzymic studies on the same sample of starch. Digestion with purified soyabean β-amylase gave an apparent conversion into maltose of 42%. Chromatographic analysis of the digest showed a trace of a second disaccharide with the chromatographic, R_f, and ionophoretic, M_G, mobilities of nigerose. After inactivation of the β-amylase and subjection of the digest to the action of R-enzyme, the β-amylolysis was increased to 52%.

In a second enzymic experiment with β-amylase, Peat et al. (1959) isolated the β-limit dextrin which had $[\alpha]_D$ + 169°, before subjecting it to the successive actions of R-enzyme and β-amylase. Analysis of the final digest showed that the dialysable sugar products included maltose, glucose, maltotriose and nigerose in diminishing order of yield. Further successive digestions of the starch with salivary α-amylase, R-enzyme and again α-amylase followed by dialysis of the digest and chromatographic analysis of the diffusate again showed nigerose together with panose, maltotriose, maltose and *iso*maltose. It was concluded that the nigerose was not an artefact and that α-1,3-links form an integral part of the polysaccharide structure.

In contrast, Barry et al. (1954) could find no evidence of uncleaved glucose after periodate oxidation of a purified sample of *Dilsea edulis* starch, and the nitrogen and sulphur contents of the thiosemicarbazide and isoniazid derivatives of the oxo-starch corresponded to a 100% α-glycol content (see p. 41) indicating the absence of 1,3-links.

Greenwood and Thomson (1961) extracted the starch from *D. edulis* [Table 1 (C)] with water at 98° for 1 hr under nitrogen, after disruption of the cell walls by immersion in liquid ammonia, treatment which is without effect on the starch. The insoluble material was removed and the solution was mixed with a saturated solution of calcium chloride and set aside at 2° for 24 hr. The precipitated calcium salts were removed, the solution dialysed and the floridean starch isolated by freeze-drying. Contaminating galactan was removed by ultracentrifugal sedimentation, and the resulting starch gave only glucose on hydrolysis.

The properties of the three samples of floridean starch (A), (B) and (C) discussed above are given in Table 1, together with those of waxy maize starch, potato amylopectin and glycogen.

Table 1   Properties of samples of floridean starch
and some other branched α-glucans.

	Floridean Starch			Waxy Maize starch	Potato amylopectin	Glycogen
	(A)	(B)	(C)			
Blue value	0·065	—	—	0·16	0·176	0·02
λ-max	530	500	550	520	550	420–490
β-Amylolysis $P_M{}^a$	42	46[c]	49	52	56	40–50
Increase in β-amylolysis after R-enzyme (see Fig. 2)	10	9[c, d]	—	12	—	0
Salivary α-amylolysis	—	65[c]	—	88	—	70
$[\eta]$ (Limiting viscosity number, see p. 50)	—	—	160	—	200	7
$[\alpha]_D$ in water	$+173°$	$+176°{}^c$	$+190°{}^b$	—	$+195°$	$+191°$
Average chainlength glucose residues	15	9[c]	18·6	20	24	10–14
Average internal chain length	—	—	7	—	8·1	6

[a] P_M = apparent conversion to maltose (see p. 44) (see Fig. 2).
[b] in 0·1M-sodium hydroxide
[c] corrected as pure glucan.
[d] after treatment with isoamylase.

Many similarities between floridean starch, higher plant amylopectins and glycogen are apparent, but the most striking difference from glycogen is the latter's immunity to R-enzyme, although this must be accepted with reserve since Fleming and Manners (1958) believe that the specificity of R-enzyme is controlled not by the degree of branching in the substrate or the exterior chain length, but by the average length of the interior chains which must not be less than five glucose units, and these authors described an eighteen unit glycogen from rabbit liver which was attacked by R-enzyme.

Recently J. R. Turvey (private communication) has examined the starches from *Ceramium rubrum*, *Gigartina stellata*, *Gracilaria confervoides*, *Porphyra umbilicalis* and *Corallina officinalis* and found that their average chain lengths are 11 to 13 with one of 15 glucose units.

Various reasons may be advanced for the variations in chain length reported for floridean starch. It has been suggested (Fleming *et al.*, 1956) that the different results, mainly based on periodate oxidation, may be due to incomplete oxidation of the polysaccharide, but it is also possible, in view of the purification difficulties, that the discrepancy is due to contamination of the starch by a galactan or by protein (Anderson and Greenwood, 1955) or that degradation during extraction may have occurred (Bottle *et al.*, 1953).

Greenwood and Thomson (1961) found a weight average molecular weight for sample (C) (from light scattering measurements) of 7×10^8. It closely resembled potato amylopectin and differed from glycogen in its variation of sedimentation coefficient with concentration, and in its iodine binding power.

Meeuse and Kreger (1954) subjected the starches from four species of *Odonthalia* to X-ray examination and from comparison of the X-ray diagram with that from potato starch decided that the *Odonthalia* extracts were true starches, but that they differed from those of higher plants in their reluctance to form colloidal solutions and their slightly higher resistance to periodate oxidation. Their blue values and β-amylolysis limits more closely resembled those of glycogen.

More recently, Meeuse and his colleagues (1960) made a survey of some thirty species of red algae growing on the Pacific Coast and found that *Constantinea subulifera* Setchell was the most suitable for the isolation of pure floridean starch. These authors extracted the starch by grinding the fresh alga in a strong iodine-potassium iodide solution and purified it by treatment for thirty minutes with butanol and boiling water. The purified product gave the same percentage of glucose as potato starch on hydrolysis with the enzyme from *Cryptochiton* and had a comparable specific rotation of $[\alpha]_D$ $+196 \cdot 5°$. All attempts to separate an amylose fraction were unsuccessful. The optical properties of the iodine complex were intermediate between those of the same complex of starch and that of glycogen. The α- and β-amylolysis limits were about 10% lower for the *Constantinea* starch than those obtained for potato starch.

Periodate oxidation under the conditions used by Fleming *et al.* (1956) of starches from *Constantinea subulifera* and *Odonthalia washingtonensis* and measurement of the formic acid released gave average chain lengths of 12 and 14, respectively, and 1 to 2% of uncleaved glucose was recovered from the oxidized polysaccharides, supporting the possible presence of a small amount of 1,3-linkages.

Summarizing the evidence from the various investigations, all these starches consist of highly ramified molecules comprising chains of α-1,4-linked glucose units interspersed very occasionally, at least in some samples, by α-1,3-linked glucose units, and all with branch points at C-6 (see Fig. 2). In many of its properties, Rhodophycean starch resembles the amylopectin of the higher plants, but in others it is intermediate between the latter and the animal polysaccharide, glycogen.

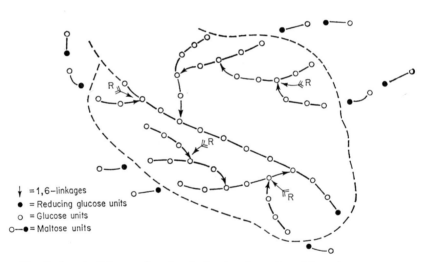

Structural Units found in floridean starch

FIG. 2. A part of the starch molecule showing the action of β-amylase — — — and of R-enzyme ≫—→.

The starch grains in these red algae, although birefringent, are smaller (*Odonthalia floccosa* 1 to 25μ), more irregular, less organized and are hydrolysed more rapidly than the grains of land plant starches (Meeuse *et al.*, 1960), although microscopic examination indicates that the hydrolytic attack proceeds in the same manner as on wheat starch grains.

B. CHLOROPHYCEAE STARCHES

The presence of starch grains in the Chlorophyceae has been accepted for many years, but it is only recently that starches in the marine species have been isolated and characterized. Meeuse and Kreger (1954, 1959) examined the X-ray diffraction patterns of a number of these starches, and they found that *Rhizoclonium* spp and *Enteropmorpha intestinalis* starches gave the A spectrum normally associated with cereal starches, that from *Codium fragile* gave the B spectrum of tuber starches and that from *Ulva expansa* closely resembled that of potato starch.

Few of the green seaweeds have been examined chemically, but those that have appeared to contain about 2% of the dry weight of a starch-type polysaccharide. The first to be isolated as a pure polysaccharide was that from *Caulerpa filiformis*. Fractionation with cetyltrimethylammonium bromide of the aqueous extract of this weed precipitated sulphated carbohydrate material and left a glucan in solution (Mackie and Percival, 1960). This was precipitated as an amorphous powder with ethanol. It had $[\alpha]_D + 154°$, gave a purple colour with iodine and contained 98% glucose. Attempted separation of an amylose fraction was unsuccessful, but subsequent studies on other genera of the Chlorophyceae revealed that the deproteinization with trichloracetic acid of the aqueous extract from *C. filiformis* before isolation of the starch would have destroyed any amylose that might have been present. The separated starch, apart from molecular size, had the characteristic properties of a typical amylopectin (Table 3). In addition, 0·98 mole of periodate was reduced for every anhydro-sugar residue and the oxopolysaccharide, isolated after dialysis, was devoid of unattacked glucose units, thus indicating their $(1 \rightarrow 4)$-linkage and the absence of 1,3-linked units (see p. 35). This was further substantiated, and the presence of 1,6-branch points confirmed by the separation of the following sugars from the hydrolysate of the methylated polysaccharide: 2,3,4,6-tetra-*O*-methylglucose (1 part), 2,3,6-tri-*O*-methylglucose (22 parts) and 2,3-di-*O*-methylglucose (1 part). These proportions correspond to an average chain length of 23, and the formic acid released on periodate oxidation gave an average chain length of 21 (cf., cereal amylopectins, average chain length 25).

The tri-*O*-methyl derivative of the *Caulerpa* amylopectin had a number average molecular weight (determined by isothermal distillation, see p. 49) of 15,120, corresponding to 76 anhydroglucose units. This value is very small for an amylopectin-type polysaccharide, but the method of extraction, purification and also methylation would undoubtedly degrade the polysaccharide, and in the native state it certainly has a larger molecule.

Aqueous extraction of *Enteromorpha compressa*, *Ulva lactuca*, *Cladophora rupestris*, *Codium fragile* and *Chaetomorpha capillaris* resulted in the separation of a complex mixture of sulphated polysaccharides, a starch-type poly-saccharide and protein from each of them. The starches were separated as the starch-iodine complex and recovered as white powders in yields of about 1% of the dry weight of the weed (Love et al., 1963). Apart from that from C. capillaris, where the amylose was destroyed during extraction, the rest were fractionated by thymol into amylose and amylopectin. The properties of these respective materials compared with the amylose and amylopectin from potato starch are given in Tables 2 and 3.

Table 2   Properties of amylose-type molecules from:

	Clado-phora rupestris	Entero-morpha compressa	Ulva lactuca	Codium fragile	Potato
% Amylose	20	22	37	16	25
$[\eta]$ (Limiting viscosity number, see p. 50)	78	44	41	36	300
$[\alpha]_D$ in water	+158°a	+177°	+161°a	+197°	+157°a
DPb	577	255	302	266	2220
Blue value	1·2	0·66	1·14	0·96	1·21
λ max	635	610	630	610	640
β-Amylolysis					
pH 3·6c P_Md	73	—	71	84	80
pH 4·6 P_M	88	76	90	96	100
Reduction $10'_4$/anhydro unit	0·97	1·08	1·01	1·10	1·01

   [a] Measured in molar KOH.
   [b] Calculated from the approximate relation DP $= 7\cdot4[\eta]$ (Cowie and Greenwood, 1957).
   [c] The β-amylase preparation was contaminated with Z-enzyme, the action of which is inhibited at pH 3·6. The algal amyloses retrograded from solution at this pH and this made measurement of the β-amylolysis difficult.
   [d] $P_M =$ apparent conversion to maltose (see p. 44).

The high positive rotation of each of the amyloses, their ready retrograda-tion from aqueous solution, their β-amylolysis limits and the reduction by each of them of about one mole of periodate for every anhydroglucose unit confirms their structure as linear $\alpha(1 \rightarrow 4)$-linked glucans (Fig. 3) and their

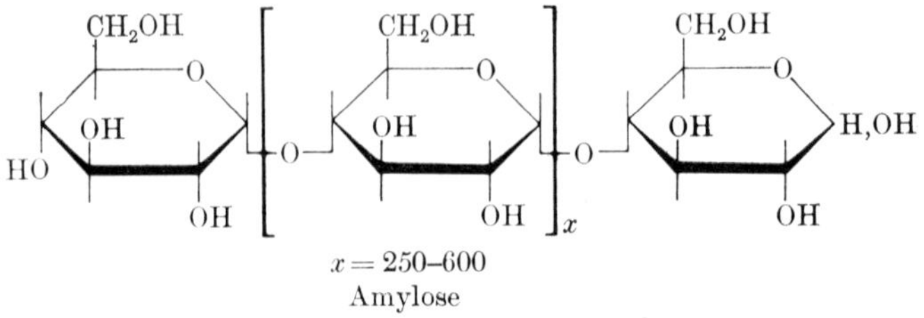

$x = 250-600$
Amylose

FIG. 3.

Table 3 Properties of amylopectin-type molecules from:

	Cladophora rupestris	Enteromorpha compressa	Ulva lactuca	Codium fragile	Chaetomorpha capillaris	Caulerpa filiformis	Potato
$[\alpha]_D$	+197°	+190°	+205°	+192°	—	+154°	+196°
$[\eta]$ (Limiting viscosity number, see page 50)	41	35	58	24	—	15	161
Blue value	0·196	0·140	0·220	0·220	0·108	0·068	0·176
λ max	565	550	560	560	540	540	560
α-Amylolysis P_M[a]	92	90	91	84	81	90	92
β-Amylolysis P_M[a]	50	—	51	51	—	57	53
β-Amylolysis after Z-enzyme	57	58	56	—	62	83[b]	56
Reduction mole IO$_4$/anhydro unit	1·05	0·90	1·05	1·2	1·11	0·980	1·04

[a] P_M = apparent conversion to maltose (see p. 44).
[b] β-Amylolysis after isoamylase action.

essential similarity with potato amylose. At the same time, their low iodine-binding power, which is in keeping with their low degree of polymerization, indicates that they are smaller molecules than a typical land plant amylose.

The striking similarity of the green marine algal amylopectins with that of potato is clearly evident from Table 3, and the structural unit (6a) depicted in Fig. 2 for floridean starch is also representative of the Chlorophyceae amylo-pectins. The structures of the *Cladophora* and *Enteromorpha* amylopectins were confirmed by methylations studies, and an average chain length of 27 was deduced for the latter from the proportion of the 2,3,4,6-tetra-*O*-methyl- and 2,3,6-tri-*O*-methyl-glucoses in the methylated hydrolysate.

Again the intrinsic viscosities are considerably lower than that of potato amylopectin and this probably indicates a much smaller molecule, a fact already confirmed for the *Caulerpa* polysaccharide. This may be the reason for the smallness of the algal starch grains, those of *Ulva expansa* measuring only 2 to 5μ (Meeuse and Kreger, 1954). Furthermore, the green algal starches show no appreciable birefringence, a characteristic property of other starches, they dissolve readily in boiling water and fail to yield a paste. It has also been found (Meeuse and Smith, 1962) that, like floridean starch, the starch grains from *Codium* and particularly those from *Ulva* are more readily hydrolysed than the land plant starches.

C. CYANOPHYCEAE STARCHES

Starches have been investigated in members of the Cyanophyceae which grow in freshwater (see Meeuse, 1962), but no chemical or enzymic study of a starch from a marine member of this Class has been reported.

II. FRUCTAN

Examination of four species of green marine algae belonging to the order Dasycladales has revealed the presence of fructans. In *Dasycladus vermicularis* and *Cymopolia barbata* (du Merac, 1955, 1956) these are low molecular weight materials whereas in *Acetabularia mediterranea* and *Batophora oerstedi*, higher molecular weight fructans are also present. *A. mediterranea*, commonly known as the mermaid's wine-glass, is found on the French mediterranean coasts and along the shores of the Adriatic. In 1863, C. Nägeli observed a mass of birefringent crystals in this alga and du Merac (1953) has shown that these comprise a fructan. It was extracted with boiling water and precipitated with alcohol as a white powder, $[\alpha]_D$ $-39°$. Both acidic and enzymic hydroly-sis gave fructose. On the basis of this, the rotation and the microscopic studies the author describes it as inulin. Similar extraction of *B. oerstedi* gave after purification a fructan in the form of highly refractive, small, white spheres which were soluble in hot water and reprecipitated on cooling. Different samples of weed gave yields of 6·5 to 12·5% of the dry weight (after subtraction of ash, which was mainly adhering rocky substrate) (Meeuse, 1963). The fructan has $[\alpha]_D - 40°$ and gave only fructose on hydrolysis. The latter was characterized chromatographically, by its rotation, and as the crystalline osazone. The X-ray diffraction pattern of this fructan was identical with that of a pure inulin except for an additional peripheral ring which the author attributed to impurity.

No structural studies have been carried out to determine the linkages between the fructose units in these polysaccharides, and, in the opinion of the present writers, until these have been done, their structure as 2,1'-linked polymers must be accepted with caution, particularly in view of the similarity of the rotation of the 2,6'-linked fructan, levan (see Fig. 4). It is possible that the additional ring in the X-ray pattern is indicative of more than a single type of linkage.

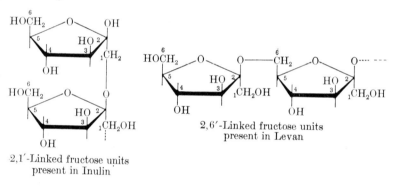

2,1'-Linked fructose units present in Inulin

2,6'-Linked fructose units present in Levan

FIG. 4.

Structural Polysaccharides

III. CELLULOSE

Cellulose is generally present in the plant cell wall as aggregates of fibrils or partly crystalline bundles which themselves consist of parallel chains of β-1,4-linked glucose residues. Where the chains are in perfect alignment, the cellulose has a crystalline structure. The cellulose fibrils are generally embedded in other polysaccharide material, the hemicelluloses in the higher plants. In algae, as will appear later, other types of polysaccharide constitute the encrusting material.

The use of cellulose in paper manufacture and for other industrial applications, has led to certain clearly defined methods of extraction of the cellulose from the plant material, and to the definition of arbitrarily determined forms of cellulose which result from these extraction procedures. For example, α-cellulose is, according to Tappi (1961), the material in cellulose pulp which is insoluble in cold 8·3% sodium hydroxide after previous swelling in 17·5% alkali, whereas β-cellulose is the material dissolved in the alkali and reprecipitated by acid. In many publications, these terms are used with slightly different or less clearly defined meanings. Often these different forms of cellulose are admixed with small amounts of other polysaccharides and therefore differ from the high molecular weight linear chains of β-1,4-linked glucose units which is the chemist's conception of cellulose.

The situation is further complicated by the presence of different physical forms of cellulose which have been recognized by X-ray studies on plant cell walls. Highly crystalline cellulose gives a characteristic well-defined X-ray diffraction pattern, is known as cellulose I, and yields only glucose on hydrolysis. It is the physical form normally found in higher plants. Treatment of cellulose with alkali alters the X-ray diffraction pattern and gives cellulose II or "mercerized" cellulose which has a lower crystallinity than cellulose I. The significance of the different forms in the studies on algal cellulose is not, at first sight, apparent, but their importance lies in the fact that the physical form of the cellulose is readily altered and changes may well occur in the native cellulose during the pre-extraction of other polysaccharide material from the alga.

Microfibrillar structures have been clearly demonstrated in electron micrographs of the cell walls of a number of algae (Preston, 1965–66; Frei and Preston, 1961), and X-ray diffraction patterns obtained which indicate the presence of cellulose, but in the absence of chemical evidence, these must be interpreted with caution. For example, the so-called α-cellulose isolated under the standard conditions from *Porphyra umbilicalis* gives a pattern closely resembling cellulose II, although it consists almost entirely of a mannan (Myers and Preston, 1959).

As a result of microscopic and X-ray examination of the cell walls, cellulose has been reported to be present in all the major Classes of marine algae, but few chemical investigations into the structure of these materials have been carried out.

A. DETERMINATION OF CELLULOSE IN SEAWEED SAMPLES (Black, 1950)

Ground seaweed (2 g) is boiled under reflux with 1·25% sulphuric acid (200 ml) for 30 min, filtered through a sintered glass crucible, washed free from acid and the residue boiled with 1·25% sodium hydroxide (200 ml) for 30 min, and again filtered as before (or centrifuged), and washed with water. The residue is kept overnight in saturated chlorine water (100 ml), then filtered, washed free from chlorine and then treated with hot 0·1N-sodium hydroxide (50 ml). After removal of alkali by washing with water, the white residue, considered to be cellulose, is dried with alcohol and ether and heated at 50° for 10 min. Under these conditions *F. vesiculosus* gave 1·7% cellulose.

B. CELLULOSE IN THE PHAEOPHYCEAE

The presence of cellulose in brown seaweeds was postulated by a number of early workers (Kylin, 1915; Naylor and Russell-Wells, 1934; Russell-Wells, 1934), but Dillon and O'Tuama (1935) were the first to prove that a cellulose-like material, which consisted of glucose and could be converted into viscose and into chloroform soluble acetyl and methyl derivatives, could be separated from *Laminaria digitata*. This was followed by structural investigations by Percival and Ross (1948; 1949) on celluloses prepared by a standard method[1] from *Laminaria hyperborea*, *L. digitata* and *Fucus vesiculosus*. These samples gave the characteristic blue colour with iodine-zinc chloride and were readily soluble in Schweizers reagent. Glucose in a yield of 80% compared with 88% from cotton cellulose was obtained on hydrolysis, and no other reducing sugars could be detected in the hydrolysate.

Comparable fluidity measurements (Clibbens and Little, 1936) of these samples with cotton cellulose treated to the isolation procedure of the algal cellulose gave values of about 40 and 28 poises^{-1}, respectively. Initially, the cotton cellulose had a value of about 2 poises^{-1}, and considerable degradation during the extraction treatment appeared to have taken place.

Evidence of β-1,4-linkage in the algal celluloses was obtained by the isolation of cellobiose octa-acetate in 31% yield, and the presence of 1,3- and 1,6-linkages was negatived by periodate oxidation studies. Sodium periodate, 1·02 moles per anhydroglucose unit, was reduced, the expected amount for 1,4-linked residues, whereas 1,3-linked units, apart from end units, would be without action on periodate and 1,6-linked units would each reduce two moles of periodate and release a mole of formic acid (see pp. 33 to 35). In the present experiments, measurement of the formic acid released gave results which corresponded to a chain of about 160 1,4-linked units (Fig. 5).

The smallness of this value compared with that of cotton cellulose is probably due to the degradation during extraction. The X-ray pattern was the same as that of normal cellulose.

The percentage of cellulose in different species of the Phaeophyceae (Black, 1950) is given in Table 4.

[1] "Methods of Analysis", Assoc. Off. Agr. Chem., Washington, 1935.

4+

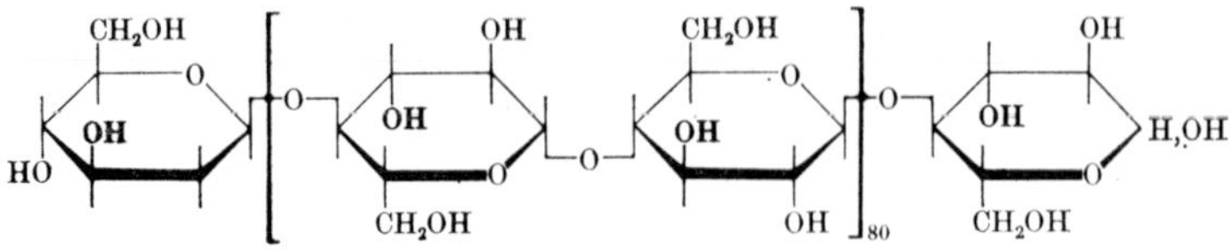

Fig. 5. Algal Cellulose.

Table 4   The cellulose content expressed as percentage of
dry weight of a number of brown algae.

	Frond	Stipe
Laminaria digitata	3–5	6–8
Laminaria saccharina	4–5	7–8
Laminaria hyperborea	4–5	8·5–10
Fucus spiralis	4·4–4·75	
Fucus serratus	2·0–3·5	
Fucus vesiculosus	1·25–2·75	
Ascophyllum nodosum	2	
Pelvetia canaliculata	0·6–1·5	

C. Cellulose in the Rhodophyceae

Few structural studies on cellulose in red algae have been recorded. From
staining reactions and solubility, cellulose has been described as a wall component of many species of red seaweed (Naylor and Russell-Wells, 1934),
although this has been disputed for the cell walls of *Griffithsia flosculosa*
(Myers *et al.*, 1956), and Cronshaw *et al.*, (1958) were unable to find any glucose
containing polymer in *Porphyra* sp.

A. G. Ross (1953) carried out a wide survey of red seaweeds, and measured,
as a percentage of the dry weight, the residual fibre, which he called cellulose, after successive hot acid and alkaline extraction. A list of the species
examined with their cellulose contents is given in Table 5.

Todokoro and his associates (Todokoro and Yoshimura, 1935; Todokoro
and Takasugi, 1936) reported that the cellulose of *Iridaea laminarioides*
consisted of 91% α-cellulose. Other Japanese workers (Araki and Hashi,
1948) reported the presence of 12·5% of cellulose in *Gelidium amansii*. This

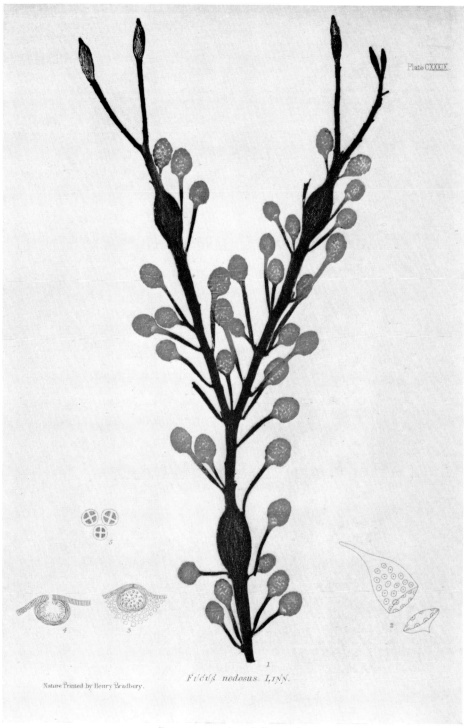

Plate CXXXIX.

5

4   3

2

1

Fucus nodosus LINN.

Nature Printed by Henry Bradbury.

Present name *Ascophyllum nodosum.*

Plate CXXXVI

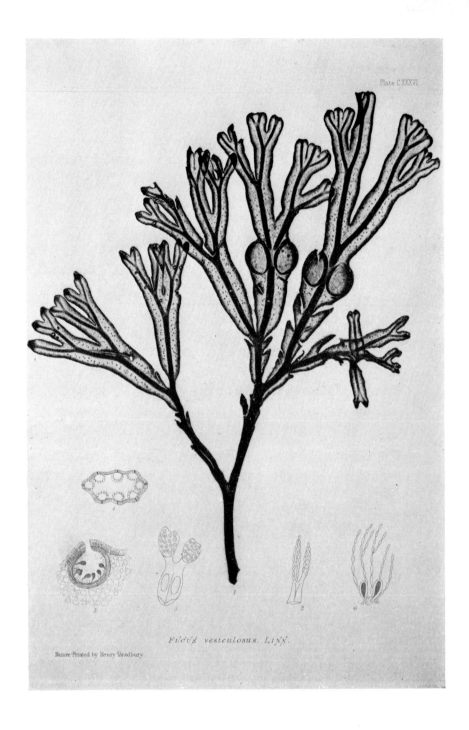

Fucus vesiculosus. LINN.

Nature Printed by Henry Bradbury

Facing page 87

Table 5.

Species	Cellulose %	Species	Cellulose %
Corallina officinalis	5·1	Ceramium rubrum	4·7
Rhodochorton floridulum	7·6	Dilsea edulis	3·1
Ahnfeltia spp.	8·8	Endocladia muricata	1·0
Rhodymenia palmata	2·4	Polysiphonia fastigiata	1·2
Gelidium pristoides	6·5	Membranoptera spp.	4·4
Gelidium cartilagineum	9·0	Phycodrys spp.	3·4
Gelidium coulteri	8·5	Furcellaria fastigiata	5·7
Suhria vittata	7·9	Gigartina stellata	2·3
Gracilaria confervoides	3·8	Gigartina radula	1·8
Gracilaria foliifera	4·6	Gigartina stiriata	1·1
Pterocladia pyramidale	8·2	Gigartina cristata	4·1
Ptilota plumosa	4·7	Chondrus crispus	2·0
Porphyra umbilicalis	3·2	Rhodoglossum affine	3·0

material more closely resembles Koliang cellullose than wood cellulose in particular in its β-cellulose content of about 18·5% (cf., wood cellulose, 6%; Koliang cellulose, 15·7%). The separated α-cellulose from *G. amansii* gave the same yield of triacetyl cellulose, octa-acetyl cellobiose and D-glucose as cotton cellulose. It differed somewhat from the latter in its colloidal properties.

X-ray diagrams of Rhodophyceae examined by Frei and Preston (1961) are consistent with the presence of cellulose I as the skeletal substance except in the Bangiales. The content and the degree of crystallinity are both rather low. The microfibrils contain a range of sugar residues other than glucose units, but it has been shown for one species, at least, that the central core of the microfibril consists solely of cellulose I (Dennis and Preston, 1961).

D. Cellulose in the Chlorophyceae

X-ray diagrams indicate the presence of cellulose I in a variety of families in the Chlorophyceae, but in many instances this is encrusted in polysaccharides comprising uronic acids and a wide variety of neutral sugars predominantly galactose, arabinose and xylose. All attempts to separate a pure cellulose (McKinnell and Percival, 1962; Cronshaw et al., 1958) from this Class of algae have been unsuccessful and chemical investigation of the fine structure is still awaited.

IV. XYLANS

A. FROM THE PHAEOPHYCEAE

Xylose has been reported as a constituent of many Phaeophyceae, and Dewar (1954) reports the preparation of a crystalline derivative of xylose from a hydrolysate of *F. vesiculosus* weed from which low molecular weight carbohydrates had been removed. Furthermore Lloyd (1960) in the purification of fucoidan from *F. vesiculosus* on DEAE-cellulose separated a small quantity (54 mg) of a practically pure xylan. In a second larger scale separation, however, the xylan fraction was found to contain some uronic acid residues. Recent investigations on *Ascophyllan nodosum* have led to the separation of xylose containing polysaccharides which also comprise glucuronic acid and fucose residues (see p. 176). The separation of a pure xylan containing only xylose units from this Class of algae has not so far been achieved.

B. FROM THE RHODOPHYCEAE

A pure xylan, $[\alpha]_D - 99°$ (in water) has been extracted with water from *Rhodymenia palmata* ("dulse") (Percival and Chanda, 1950; Barry *et al.*, 1950), where it constitutes the major polysaccharide. It is included in this section for convenience; although its solubility in water indicates that it is probably a food reserve polysaccharide, direct evidence on this point is lacking.

Methylation, followed by hydrolysis, afforded 2,3,4-tri-, 2,4-di-, 2,3-di- and 2-*O*-methyl-D-xyloses. This suggests branched molecules with chains containing both 1,3- and 1,4-linkages. Periodate oxidation resulted in cleavage of 80% of the units which must therefore be 1,4-linked. Methylation analysis indicated an average chain length of 20 to 21, and the DP determined from the formaldehyde released on periodate oxidation (see p. 35) of another sample gave a molecular size of 39 to 40 xylose units. It is therefore an

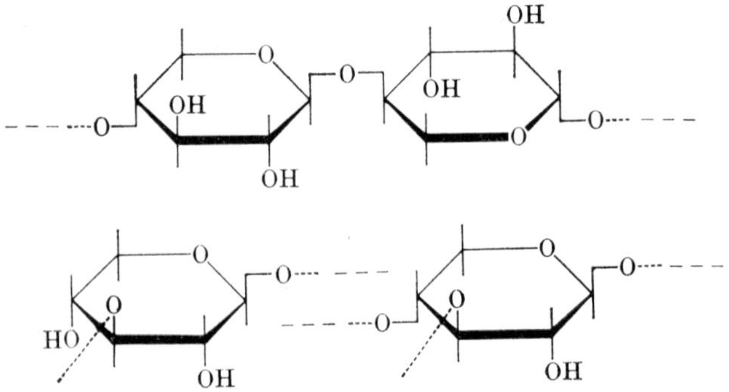

FIG. 6. Structural units present in *Rhodymenia* xylan.

essentially linear molecule of about 40 xylose units with, on average, one branch point in each molecule (Fig. 6).

That the two types of linkage are present in a single molecule was shown by hydrolysis by a bacterium from sheep rumen (Howard, 1957). A trisaccharide, O-β-D-xylopyranosyl(1 $\rightarrow$ 3)-β-D-xylopyranosyl(1 $\rightarrow$ 4)-D-xylopyranose (3^2-β-xylosylxylobiose) (7a), and a tetrasaccharide, O-β-D-xylopyranosyl(1 $\rightarrow$ 4)β-D-xylopyranosyl(1 $\rightarrow$ 3)β-D-xylopyranosyl(1 $\rightarrow$ 4)-D-xylopyranose (7) were separated from the enzymic hydrolysate.

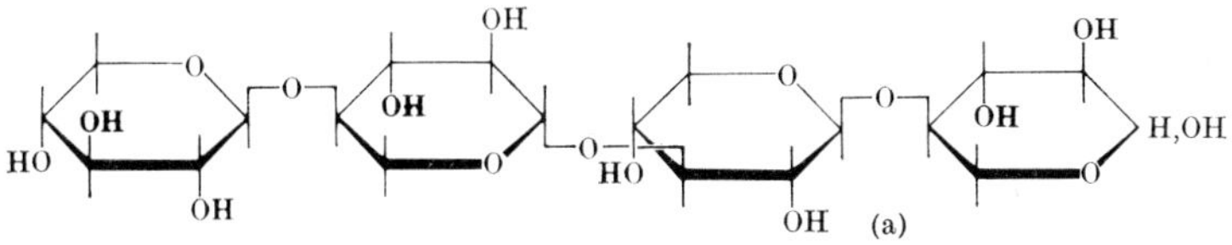

OH
OH
OH
OH
OH
OH
HO
HO
O
O
O
O
H,OH
OH
OH (a)
(7)

Support for these findings was obtained by the application of Barry degradation (Barry et al., 1954) (see p. 41) and of Smith degradation (Manners and Mitchell, 1963). The latter resulted in the formation of xylose, xylitol, glycerol, xylosylglycerol and a small amount of rhodymenabiose, O-β-D-xylopyranosyl (1 $\rightarrow$ 3)-D-xylopyranose (9). In addition, these results may be interpreted as showing the presence of only a small proportion of adjacent 1,3-linked units, the majority being flanked by 1,4-linked xylose residues. This was confirmed by degradation with an enzyme preparation from *Myrothecium verrucaria* which hydrolysed the xylan to xylose, xylobiose (8), -triose, -tetraose, and small amounts of rhodymenabiose (9) and 3^2-β-xylosylxylobiose (7a) together with trace amounts of two trisaccharides (see also Manners and Mitchell, 1967).

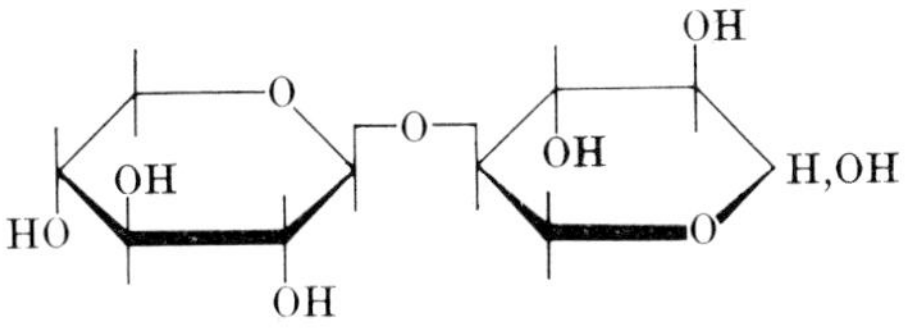

OH
OH
OH
HO
O
O
O
H,OH
OH
(8)

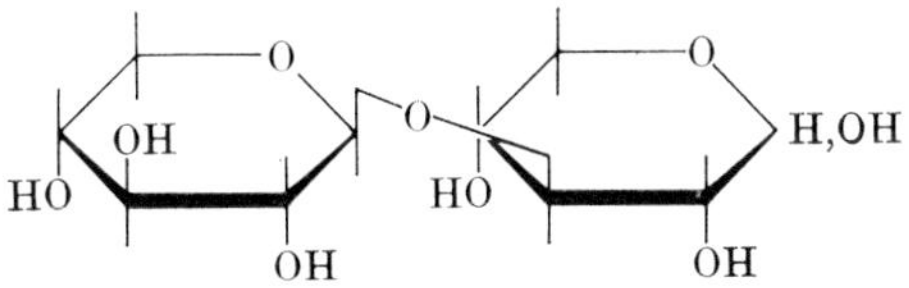

OH
OH
HO
HO
O
O
O
H,OH
OH
OH
(9)

Further enzymic studies on *Rhodymenia* xylan with a purified enzyme from *Stereum sangvirolentum* (Bjorndal *et al.*, 1965) which in addition to hydrolysing β-1,4-linked xylans had strong cellulase activity, enabled further deductions on the relative positions of the 1,4- and 1,3-linkages in this xylan to be made. After incubation of the polymer with the enzyme for 28 days, when constant rotation had been attained, the solution contained a hexasaccharide, two pentasaccharides and two tetrasaccharides all of which contained one β-(1 → 3)-linkage. From the proportions and structures of these oligosaccharides it was concluded that the enzyme preferentially attacks β-1,4-linkages which are flanked by other 1,4-linkages and that the β-1,3-linkage not only inhibits the cleavage of an adjacent unit, but also reduces the rate of cleavage of a linkage two-removed. From these deductions it was concluded that a regular arrangement of the 1,3- and 1,4-linkages does not occur in the macromolecule, but that a purely random sequence is the more probable.

Recent work (Bjorndal *et al.*, 1965) has also revealed that aqueous and dilute acid extraction of *R. palmata* yields two xylans (I and II) both of which methylation studies showed to be essentially linear structures containing β(1 → 3)- and β(1 → 4)-linkages. However, the proportions of the two linkages in the two polymers are different. Periodate oxidation confirmed this finding and it was calculated that xylan I contains 29% of β(1 → 3) and 71% of β(1 → 4) whereas xylan II comprises 38% of the former and 62% of the latter. The DP of xylan I (determined osmometrically on the nitrate) was found to be 64, and this was in reasonable agreement with the figures calculated from the formic acid released on periodate oxidation of the xylan before and after reduction. Xylan II gave an insoluble nitrate, but the DP from formic acid release was of the same order as that of xylan I.

Previous determinations by Manners and Mitchell (1963) of the formic acid produced following periodate oxidation under conditions which avoid "over-oxidation" (see p. 36), on two xylans of similar constitution (one prepared by acid extraction and the other by extraction with butanol-water mixtures of *R. palmata*) gave DPs of 54 and 114, respectively. The latter authors considered that degradation had occurred during the acid extraction, but in view of the later investigations, it seems that although the basic structure is the same for all the *Rhodymenia* xylans, the proportion of the two types of linkage may vary in the different molecules.

Frei and Preston (1964b) found it possible to mechanically separate the cell wall proper and the cuticle of *Porphyra umbilicalis*, and were able to show that the latter consisted of a β-1,4-linked mannan (see p. 93) and that the cell wall consisted of a xylan. X-ray and electron microscopic examination of the wall showed that the polysaccharide was arranged in a random net-work of microfibrils with indications of a crossed microfibrillar pattern very similar to the arrangement found in the xylans which comprise the cell walls of some of the Siphoneous green algae (see below), and from this these authors conclude that the *Porphyra* xylan is a β-1,3-linked polymer. Similar examination of *Bangia fuscopurpurea*, a marine alga closely related to *Porphyra*, indicated that the cell walls are also constituted of a β-1,3-linked xylan. Nevertheless until these latter two xylans have been subjected to chemical investigation, the presence of 1,4-linked units cannot be excluded.

Recent studies (Bowker, 1966) have indicated the presence of an alkali soluble xylan in *Laurencia pinnatifida*.

C. From the Chlorophyceae

The cell walls of a number of green algae consist mainly of xylan, although a small proportion of polymeric material comprising other sugars may also be present in the wall.

Bryopsis maxima, *Caulerpa brachypus*, *Caulerpa racemosa*, and *Halimeda cuneata* were collected from the water of the Pacific coast of Central Japan, *Udotea orientalis*, *Chlorodesmis formosana* and *Pseudodichotomosiphon* from latitude 26·5 N on the Okinawa island, *Caulerpa filiformis* from the rock pools near Cape Town and *C. racemosa* and *C. sertularoides* from dead coral in strong surf on the Gata Islands, and all were shown to have similar cell walls.

From the Japanese samples, the "crude fibre" which constituted the cell walls was prepared (Miwa *et al.*, 1961; Nisizawa *et al.*, 1960) by treatment of the weed with cold acid, followed successively by hot 1·25% sodium hydroxide and 1·25% sulphuric acid. Although the residue was washed with water and bleached with sodium chlorite, the derived crude fibres retained the original membrane structure. They showed neither the staining reactions nor optical anisotropy given by cellulose, apart from the *Bryopsis* fibre which stained blue violet with iodine-potassium iodide-zinc chloride. In addition to 76 to 79% of xylose, small amounts of glucose were invariably present in the hydrolysates of these fibres and other sugars were also reported from some species.

The xylan from *C. filiformis*, purified by precipitation as the copper complex, was found to retain about 10% of glucose units and had $[\alpha]_D - 31°$ (Mackie and Percival, 1959). Dialysis against running water for 4 days reduced the glucose content to about 4% and analysis gave a xylose content of 96%. Methylation and hydrolysis gave 2,3,4-tri-*O*-methyl-, 2,4-di-*O*-methyl- and mono-*O*-methyl-xyloses and 2,4,6-tri-*O*-methylglucose in the approximate molar ratios of 2:96:1:0·5, respectively, together with trace quantities of 2,3,4,6-tetra-*O*-methyl- and 4,6-di-*O*-methyl-glucoses. It may be concluded that this xylan, unlike the *Rhodymenia* xylan, is devoid of 1,4-linked units and consists of essentially linear chains of about 47 β-1,3-linked xylose units

(Fig. 7) and that the glucose units appear to be derived from a contaminating laminaran-type polysaccharide (see p. 53). This conclusion was supported by the recovery in about 70% yield of a xylan, $[\alpha]_D - 36°$, devoid of glucose units, after exhaustive extraction of the xylan, $[\alpha]_D - 31°$ with water under reflux.

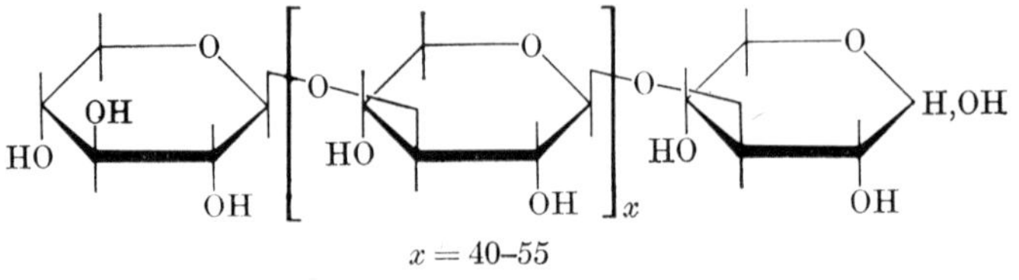

$$x = 40\text{-}55$$

Fig. 7. Chlorophyceae xylan.

Measurement of the amount of periodate oxidized by the xylan and the quantity of formic acid released gave an average chain length of 42 and 43 xylose units, respectively. In addition, the recovery of 90% of the xylose on hydrolysis of the oxopolysaccharide provides further proof of the 1,3-xylosidic linkage.

Preliminary investigation of *C. racemosa* and *C. sertularoides* xylans indicated a similar type of structure, although the latter contained something of the order of 30% of glucose. In support of the cell wall nature of these polysaccharides is the fact that the weed residue remaining after extraction of the xylan still comprised mainly xylose and small amounts of glucose.

Four of the Japanese crude fibres, those from *B. maxima*, *C. brachypus*, *H. cuneata* and *C. formosana* were extracted with 10% sodium hydroxide at room temperature and the crude xylans precipitated from the alkaline solution with ethanol. Purification was effected by dissolution in alkali and reprecipitation with alcohol followed by treatment with hot water to remove soluble impurities. The resulting xylans had specific rotations ranging from $- 26°$ to $- 38°$, and glucose contents of about 9%. By repeated extraction with hot water, the glucose content of *Bryopsis* xylan was reduced from 8·4 to 4·5%. The four xylans were very resistant to periodate oxidation and comparable amounts of periodate were reduced and formic acid released to those found for *C. filiformis* xylan. In the Japanese experiments, the calculated DPs for the xylans from these results were *B. maxima*, 45; *C. brachypus*, 48; *H. cuneata*, 45 and *C. formosana* 67. Again confirmation of the 1,3-linkage was obtained from the uncleaved xylose and glucose in the oxopolysaccharide. Confirmation of this structure for the xylan from *C. cuneata* was obtained by its hydrolysis to xylose and a small amount of glucose by a purified exo-β-1,3-xylanase from *Chaetomium globosum* A2 (Fukui *et al.*, 1960). This enzyme had no hydrolytic activity towards β-1,4-xylan, cellulose or starch.

Support for the molecular size was obtained by sedimentation experiments with $C.$ $brachypus$ xylan, a value of $7 \cdot 2 \times 10^{-4}$ being found for K_m, see p. 50. This gave a DP of 42 units. This value of K_m was used to calculate the DP (where $\eta_{sp}/c = K_m M$) of $Bryopsis$, $Halimeda$ and $Chlorodesmis$ xylans from viscosity determinations and values of 42, 42 and 54, respectively, were obtained.

The walls of this group of plants are finely lamellated and can therefore readily be prepared in a form suitable for electron microscopy. This has revealed negatively birefringent microfibrils similar to the microfibrils of cellulose, and Frei and Preston (1964a) consider that the laminaran-type glucan forms an incrusting substance in which the microfibrils are embedded. The X-ray diagrams and optical properties of the microfibrils suggest that the xylan chains are coiled helically within them.

V. MANNANS

Pure mannans have been found in the Rhodophyceae and Chlorophyceae, and a sulphated glucuronosylmannan (see p. 186) in the Bacillariophyceae. They are generally regarded as structural polysaccharides.

A. FROM THE RHODOPHYCEAE

Although a small proportion of the mannan of $Porphyra$ $umbilicalis$ is extracted by cold water and more by dilute alkali, the largest amount is obtained by hot concentrated alkali. It is considered that the different extracts are structurally similar and differ only in their molecular size or shape.

A mannan was isolated from $P.$ $umbilicalis$ with hot 20% alkali, after exhaustive extraction of the weed with water and cold dilute alkali. This was purified from contaminating xylan by the formation of the insoluble copper complex with Fehling's solution (Jones, 1950). Treatment with ethanolic hydrogen chloride liberated the free mannan, $[\alpha]_D - 41° \rightarrow -22°$ (in formic acid) as a white fibrous material (3·8% of the dried weed) which had become insoluble in hot concentrated alkali. Methylation and hydrolysis gave 2,3,6-tri-O-methyl- and 2,3,4,6-tetra-O-methyl-D-mannoses, the proportion of the latter corresponding to a chain length of about 12 units. This indicates linear chains of about 12 1,4-linked β-D-mannose units for this mannan. (Fig. 8), the β-linkage being inferred from the low specific rotation, and reveals the essential similarity of this material with the ivory nut mannan. Evidence for the absence of any 1,3-linkages and branch points was obtained from periodate oxidation as all the sugar units were cleaved by the periodate.

The insolubility of a molecule of this size can best be explained by the close alignment of the molecules to form aggregates which are possibly held together

4*

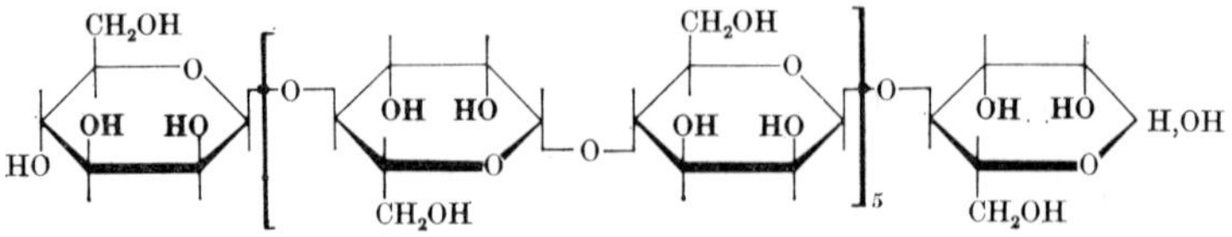

Fig. 8.

by hydrogen bonding. This mannan is considered to form the so-called structural cuticle round the cells in *P. umbilicalis*, since it is the sole constituent of the residue remaining after exhaustive extraction, which still retains the sheath-like character of the original, broken-up thallus. Mechanical separation of the cuticle supported these conclusions (Frei and Preston, 1964b). In fact, a conspicuous feature of *Porphyra* lies in the abundance of this extracellular skeletal mannan. Preston considers it unlikely that this preceded the development of the cell walls and therefore it must be produced from a precursor continually synthesized by the cytoplasm, and in the form of the monomer, dimer or even longer unit passed progressively through the wall and that it does not, beyond certain limits, accumulate within the wall itself being presumably prohibited by an interlinkage between the xylan microfibrils (see p. 93). X-ray diffraction diagrams indicated that all the manan is not crystalline and that the crystallites lie more or less at random. Examination with the electron microscope shows a granular appearance. Similar conclusions were reached for the mannan from *Bangia fuscopurpurea*.

Mannanase activity, on ivory nut mannan as substrate, has been detected in an enzymic extract from *P. umbilicalis* (Peat and Rees, 1961).

B. From the Chlorophyceae

Mannans have been isolated from members of the Chlorophyceae growing in Japanese waters, from various species of *Codium*, and from *Derbesia lamourouxii*, all members of the order Siphonales, and from *Acetabularia calyculus* and *Halicoryne wrightii*, both members of the order Dasycladales. From each of these species "crude fibre" which is considered to represent the main cell wall material was left after acid and alkaline extraction and mild chlorite treatment. Microscopic examination showed that the residue retained the almost unchanged membrane structure of the original material. All the samples gave mannose on hydrolysis and in addition the members of the Siphonales gave a small proportion of glucose, although these extracts gave none of the staining reactions of cellulose (Miwa *et al.*, 1961). Like the mannan from *P. umbilicalis*, the fibres were very insoluble and therefore difficult to purify. However extraction with 50% zinc chloride, followed by precipitation with acetone, gave a pure mannan, $[\alpha]_D - 46°$ to $- 50°$ depending on the species. The small amount of residue from the fibres was invariably optically isotropic and failed to give any staining reactions for cellulose.

Alternatively, it was found that after exhaustive aqueous extraction of samples of *Codium fragile* (from South Africa and from the South West of France), mild chlorite and dilute alkaline extraction, the residue on extraction with 20% sodium hydroxide solution at 80° under nitrogen yielded a solution of the mannan (Love and Percival, 1964), which was separated as the copper complex with Fehling's solution. Decomposition of the complex with 1% ethanolic hydrogen chloride gave a mannan, $[\alpha]_D - 41°$, comprising about 95% of mannose and 5% of glucose residues and which was insoluble in 20% alkali and very resistant to hydrolysis with acid.

Enzymic hydrolysis of this latter material with commercial hemicellulase yielded, in addition to mannose, manno-biose (4-*O*-β-D-mannopyranosyl-D-mannose), -triose, tetraose, (**10**), and a small quantity of higher oligosaccharides, a trace quantity of a disaccharide which gave mannose and glucose

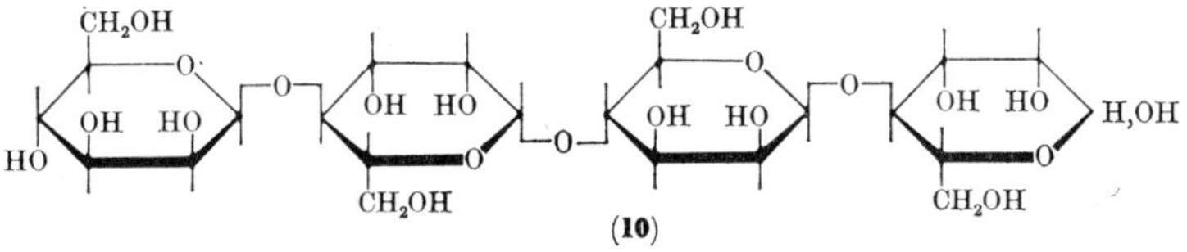

(**10**)

on hydrolysis. Acetolysis led to the isolation of the same oligosaccharides and a third disaccharide which was tentatively identified as mannose (1 → 4) glucose (**11**). Similarly, the Japanese mannans were partially hydrolysed to

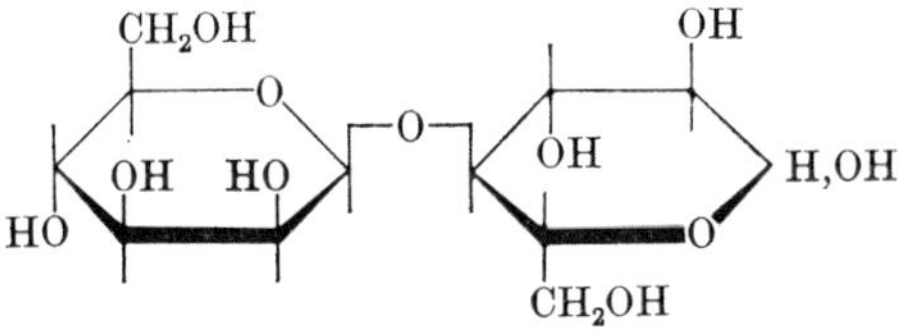

(**11**)

mannose and a mannodisaccharide by enzymes in Takadiastase and in the midgut gland of *Procambarus clarkei* and *Dolabella* sp. These results indicate the similarity of the Chlorophycean mannans to that of *P. umbilicalis* and the higher plant mannans, namely chains of β-1,4-linked mannose units (see Fig. 9). The Japanese mannans reduced 0·96 mole of periodate per anhydrosugar unit. In contrast, the mannans extracted by the Edinburgh School reduced only 0·87 mole per C_6 unit, but in view of the enzymic and acetolysis studies it is considered that the low reduction was probably due to the high insolubility of the mannan rather than to the presence of branch points or anomalous linkages. The Japanese workers measured the formic acid released on periodate oxidation of the mannan, and on the assumption of a linear chain molecule and the absence of "overoxidation" (see p. 36) decided that two moles of formic acid were produced from the ends of each chain. From their results, they calculated that the average DP was 16 (Fig. 9).

X-ray diagrams of *Codium* mannan (Preston, 1965) reveal the presence of chain molecules lying parallel to each other in the wall as in cellulose, but no microfibrils are visible in the electron microscope and it is considered that the chains are possibly too short to form microfibrils, but that the neighbouring chains overlap from one crystallite to another and so give linear coherence.

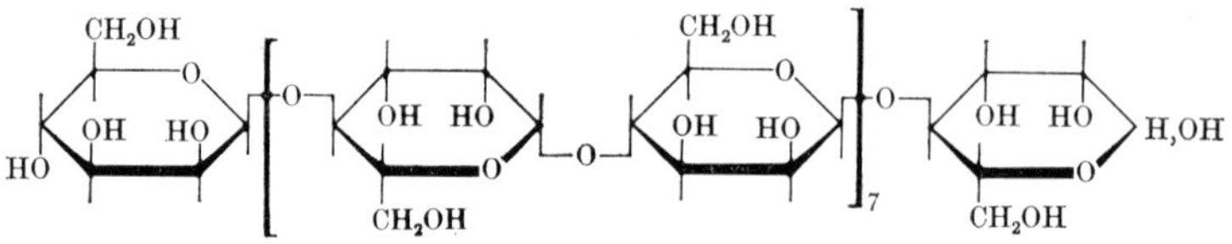

FIG. 9.

A xylomannan has recently been extracted from *Prasiola Japonica* Yatabe (H. Takeda *et al.*, private communication), a green seaweed belonging to the Schizogoniaceae.

REFERENCES

Anderson, D. M. W., and Greenwood, C. T. (1955). *J. Sci. Fd. Agric.* **6**, 587.
Araki, C., and Hashi, Y. (1948). *Memoirs 45th Anniv. Kyoto Indust. Tech. School.* p. 64. (In Japanese.)
Barry, V. C., Dillon, T., Hawkins, B., and O'Colla, P. (1950). *Nature, Lond.* **166**, 788.
Barry, V. C., McCormick, J. E., and Mitchell, P. W. D. (1954). *J. chem. Soc.* p. 3692.
Bjorndal, H., Eriksson, K. E., Garegg, Per J., Lindberg, B., and Swan, B. (1965). *Acta chem. scand.* **19**, 2309.
Black, W. A. P. (1950). *J. marine biol. Ass. U.K.* **29**, 379.

Bottle, R. T., Gilbert, G. A., Greenwood, C. T., and Saad, K. N. (1953). *Chemy. Ind.* p. 541.

Bowker, D. M. (1966). Ph.D. Thesis, University of Wales.

Clibbens, D. A., and Little, A. H. (1936). *J. Text. Inst.* **27**, T 285.

Cowie, J. M. G., and Greenwood, C. T. (1957). *J. chem. Soc.* p. 2658.

Cronshaw, J., Myers, A., and Preston, R. D. (1958). *Biochim. biophys. Acta* **27**, 89.

Dennis, D. T., and Preston, R. D. (1961). *Nature, Lond.* **191**, 667.

Dewar, E. T. (1954). *Chemy. Ind.* p. 785.

Dillon, T., and O'Tuama, T. (1935). *Sci. Proc. R. Dubl. Soc.* **21**, 147.

Fleming, I. D., and Manners, D. J. (1958). *Chemy. Ind.* p. 831.

Fleming, I. D., Hirst, E. L., and Manners, D. J. (1956). *J. chem. Soc.* p. 2831 and ref. cited therein.

Frei, E., and Preston, R. D. (1961). *Nature, Lond.* **192**, 939.

Frei, E., and Preston, R. D. (1964a). *Proc. R. Soc.* B**160**, 293.

Frei, E., and Preston, R. D. (1964b). *Proc. R. Soc.* B**160**, 314.

Fukui, S., Suzuki, T., Kitahara, K., and Miwa, T. (1960). *J. Gen. App. Microbiol.* **6**, 270.

Greenwood, C. T., and Thomson, J., (1961). *J. chem. Soc.*, p. 1534.

Howard, B. H. (1957). *Biochem. J.* **67**, 643.

Jones, J. K. N. (1950). *J. chem. Soc.* p. 3292.

Kylin, H. (1913). *Hoppe-Seyler's Z. physiol. Chem.* **83**, 171.

Kylin, H. (1915). *Hoppe-Seyler's Z. physiol. Chem.* **94**, 357.

Lloyd, K. O. (1960). Ph.D. Thesis, Wales.

Love, J., and Percival, Elizabeth (1964). *J. chem. Soc.* p. 3345.

Love, J., Mackie, W., McKinnell, J. P., and Percival Elizabeth, (1963). *J. chem. Soc.* p. 4177.

Mackie, I. M., and Percival, Elizabeth (1959). *J. chem. Soc.* p. 1151.

Mackie, I. M., and Percival, Elizabeth (1960). *J. chem. Soc.* p. 2381.

McKinnell, J. P., and Percival, Elizabeth (1962). *J. chem. Soc.* p. 3141.

Manners, D. J., and Mitchell, J. P. (1963). *Biochem. J.*, **89**, 92P.

Manners, D. J., and Mitchell, J. P. (1967). *Proc. Biochem. Soc.* March p. 15.

Meeuse, B. J. D. (1962). *In* "Physiology and Biochemistry of Algae" (R. A. Lewin, ed.), p. 291, Academic Press, New York and London.

Meeuse, B. J. D. (1963). *Acta bot. neerl.* **12**, 315.

Meeuse, B. J. D., and Kreger, D. R. (1954). *Biochim. biophys. Acta* **13**, 593.

Meeuse, B. J. D., and Kreger, D. R. (1959). *Biochim. biophys. Acta* **35**, 26.

Meeuse, B. J. D., and Smith, B. N. (1962). *Planta* **57**, 624.

Meeuse, B. J. D., Andries, M., and Wood, J. A. (1960). *J. exp. Bot.* **11**, 129.

Mérac, M. L. Rubat du. (1953). *Revue gén. Bot.* **60**, 689.

Mérac, M. L. Rubat du. (1955). *C. r. hebd. Séanc. Acad. Sci., Paris* **241**, 88.

Mérac, M. L. Rubat du. (1956). *C. r. hebd. Séanc. Acad. Sci., Paris* **243**, 714.

Miwa, T., Iriki, Y., Suzuki, T. (1961). *Colloques int. Cent. natn. Rech. scient.* **103**, 135.

Myers, A., and Preston, R. D. (1959). *Proc. R. Soc.* B**150**, 456.

Myers, A., Preston, R. D., and Ripley, G. W. (1956). *Proc. R. Soc.* B**144**, 450.

Naylor, G. L., and Russell-Wells, B. (1934). *Ann. Bot.* **48**, 635.

Nisizawa, K., Miwa, T., Iriki, Y., and Suzuki, T. (1960). *Nature, Lond.* **187**, 82.

O'Colla, P. (1953). *Proc. R. Ir. Acad.* **55**B, 321, and ref. cited therein.

Peat, S., and Rees, D. A. (1961). *Biochem. J.* **79**, 7.

Peat, S., Turvey, J. R., and Evans, J. M. (1959). *J. chem. Soc.* p. 3223.

Percival, E. G. V., and Ross, A. G. (1948). *Nature, Lond.* **162**, 895.

Percival, E. G. V., and Ross, A. G. (1949). *J. chem. Soc.* p. 3041.

Percival, E. G. V., and Chanda, S. K. (1950). *Nature, Lond.* **166**, 787.

Preston, R. D. (1965–66). *Advmt. Sci., Lond.* **22**, 1.

Ross, A. G. (1953). *J. Sci. Fd. Agric.* **4**, 333.

Russell-Wells, B. (1934). *Nature, Lond.* **133**, 651.

Steiner, E. T., and Guthrie, J. D. (1944). *Ind. Engng. Chem. (Analyt.)* **16**, 736.

Tappi, (1961). Standard Method, T 203.

Takeda, H., Nisizawa, K., and Miwa, T. (1966). Private communication.

Todokoro, T., and Yoshimura, K. (1935). *J. chem. Soc. Japan* **56**, 655.

Todokoro, T., and Takasugi, N. (1936). *J. Agric. chem. Soc. Japan* **12**, 421.

Alginic Acid

I. INTRODUCTION

Alginic acid, a polyuronide found in brown seaweeds (Phaeophyceae), was first obtained by E. C. C. Stanford, a chemist who was concerned with the better utilization of seaweed than burning it for the recovery of iodine. He took out a patent for his process in 1881, and made a number of suggestions (Stanford, 1883) for the commercial use of the product which he called algin, but none was successful. Small amounts of crude material were made for boiler water treatment from Stanford's time, but the modern industry dates from about 1930 with the development of production and uses in the United States and Britain. The scale of manufacture has greatly increased since the Second World War, and although production figures are not published, there is little doubt that over 10,000 tons are produced annually. The principal producing countries are the United States, Britain, France, Norway and Japan.

A. Sources

Alginic acid has been found in all of the large number of species of brown seaweed that have been examined, but only a few which can be obtained at suitable centres in large quantity are used commercially. The most important of them are *Macrocystis pyrifera* (Pacific coast of America), *Ascophyllum nodosum* (Europe), *Laminaria* species (Europe and Japan), *Ecklonia* species (South Africa).

The alginic acid content of a number of species, including those of commercial interest, has been extensively studied (Black, 1953; Steiner and Mc-Neely, 1954). Alginic acid may constitute 14 to 40% of the dry solids of these seaweeds. The proportions of the different substances making up the solid matter in the seaweeds undergo seasonal variation, and in general, the alginic acid forms a smaller proportion in the periods of rapid growth than in the colder months when little growth is taking place (Black, 1950). These figures give an exaggerated idea of the way that the amount of alginic acid varies in the plant, as when the proportion is low, the total dry solids content is high and vice versa (see p. 8).

B. STATE OF COMBINATION IN THE PLANT

The salts of alginic acid undergo base exchange reactions, and quickly come into equilibrium with salt solutions in contact with them. The alginate in the seaweed behaves as a base exchange material in the same way as an isolated alginate (Wassermann, 1948, 1949), and from a knowledge of the composition of the salts in the seaweed, it can be concluded that the alginate is present as a mixed salt with sufficient calcium to render it insoluble. It should be noted that fucoidan is also combined with bases, and there may possibly be some cross-linking of fucoidan to alginate through divalent ions.

II. ISOLATION

Although the alginate cannot be isolated from seaweed without some chemical changes, methods are available by which it is probable that there is little or no breakdown of the alginic acid molecule, and the reactions used only change the cations associated with it. Using base exchange reactions, the alginate is made alternately insoluble and soluble to enable it to be separated from the other constituents of the alga. As large molecules have to diffuse from the plant tissues, the seaweed, whether freshly collected or dried, is preferably reduced to small particles as a preliminary step.

In the first stages of the preparation the alginate remains insoluble and water soluble constituents such as salts, mannitol, laminarin, fucoidan and some colouring matter can be removed from the weed by washing it with hot water or lime water (Rose, 1951). Treatment with dilute acid in the cold with or without the prior washing converts the alginate into the free acid and removes acid soluble substances. The alginic acid can then be converted into the soluble sodium alginate and dissolved out of the seaweed by stirring it with sodium carbonate solution. This is best done in two stages, first for 1 to 4 hr in a fairly concentrated state (for example 2 to 5% dry seaweed solids), and then a further 2 to 4 hr after considerable dilution (for example, 0·5% dry seaweed solids) to obtain a solution of low enough viscosity to allow separation of the insoluble seaweed residue.

Smidsrod et al. (1963) have shown that reducing substances, present in some of the Fucacae, can lead to degradation while the alginate is in an alkaline state, and that this can be overcome by adding formaldehyde to the seaweed before it is made alkaline.

Several different methods have been used to separate the alginate from other soluble substances. Addition of an alcohol (Hirst et al., 1964; Haug, 1965) precipitates sodium alginate. Adding the solution to a calcium chloride solution with good stirring precipitates calcium alginate, and similarly adding hydrochloric acid precipitates alginic acid.

Cast rods of *Laminaria hyperborea*, South
Uist, Scotland.

Loading *Ascophyllum nodosum*, North
Uist, Scotland.

In all cases there is some co-precipitation of impurities, and solution followed by reprecipitation, perhaps repeated, is necessary to obtain a pure product. Alginic acid obtained by direct precipitation or by leaching with hydrochloric acid, is washed free from chloride and then brought into solution by titrating with sodium hydroxide. Precipitation with alcohol is perhaps the most effective method of freeing the sodium alginate from colouring matter, but acid or calcium precipitation is more effective in removing the protein and fucoidan which are generally present in the original extract. The use of one method in the first precipitation and another in reprecipitation can therefore give the purest product.

The use of oxidizing substances as bleaching agents, a normal practice in the improvement of the colour of commercial products, is not recommended in the preparation of material for structural studies as they can bring about changes in the structure of alginic acid (Whistler and Schweiger, 1958).

Although details of methods used for commercial production of alginates have not been published, a number of patents on the subject have been taken out (Maass, 1959; Steiner and McNeely, 1954). All the methods are modifications of the original Stanford process.

III. CONSTITUTION AND STRUCTURE

Nelson and Cretcher (1929, 1930) were the first to establish that mannuronic acid was a constituent of alginic acid and by decarboxylation of alginic acid they showed that the released carbon dioxide (24·2%) was equivalent to a uronic acid content of 100%.

The acidity of alginic acid made complete methylation difficult, but this was achieved on a degraded sample with thallous hydroxide and methyl iodide (Hirst *et al.*, 1939). Methanolysis of the methylated material gave the methyl ester of methyl 2,3-di-*O*-methyl-D-mannuronoside (**1**) which was characterized by degradation to 2,3-di-*O*-methylerythraric acid (**2**). These results were extended to a less degraded alginic acid with chain length of about a hundred units (Chanda *et al.*, 1952). After Haworth methylation and hydrolysis with 98% formic acid, the products after conversion into the

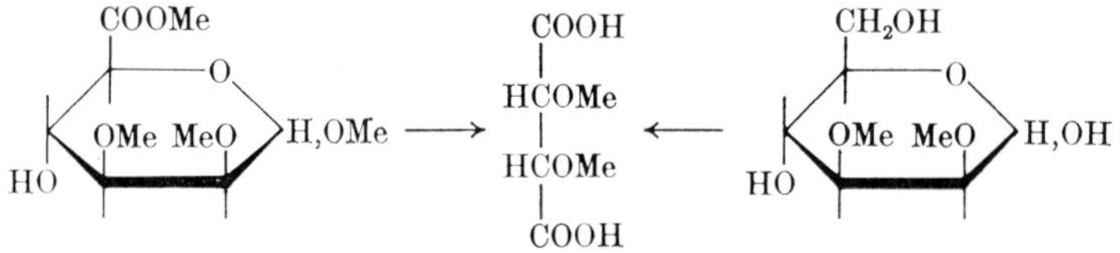

FIG. 1.

methyl ester methyl glycosides were reduced with lithium aluminium hydride and the derived methyl glycosides hydrolysed. The major component of the final hydrolysate was characterized as 2,3-di-*O*-methyl-D-mannose (**3**) by conversion into the 2,3-di-*O*-methylerythraric acid (**2**). From these results, coupled with the high negative rotation ($[\alpha]_D - 139°$), it was concluded that a high proportion of the molecule consisted of β-1,4-linked D-mannuronic acid units [Fig. 2, (**4**)]. This was confirmed by the isolation of erythraric acid [mesotartaric acid (**6**)] after periodate oxidation, bromine oxidation and hydrolysis of alginic acid (Fig. 2) (Lucas and Stewart, 1940b).

The next major advance came in 1955 when it was shown (Fischer and Dörfel, 1955) that variable amounts of L-guluronic acid [Fig. 2, (**5**)] units are also present in samples of alginic acid from seventeen different genera of brown seaweeds. Improved chromatographic techniques enabled the isolation of threaric acid [L(+)-tartaric acid (**7**)] in addition to erythraric acid when alginic acid was subjected to the above procedure of oxidation with periodate and bromine (Drummond *et al.*, 1958, 1962) or with hypochlorite (Whistler and Schweiger, 1958). This provided evidence that the guluronic acid residues weer also linked through C-1 and C-4 [Fig. 2, (**5**)]. Identification

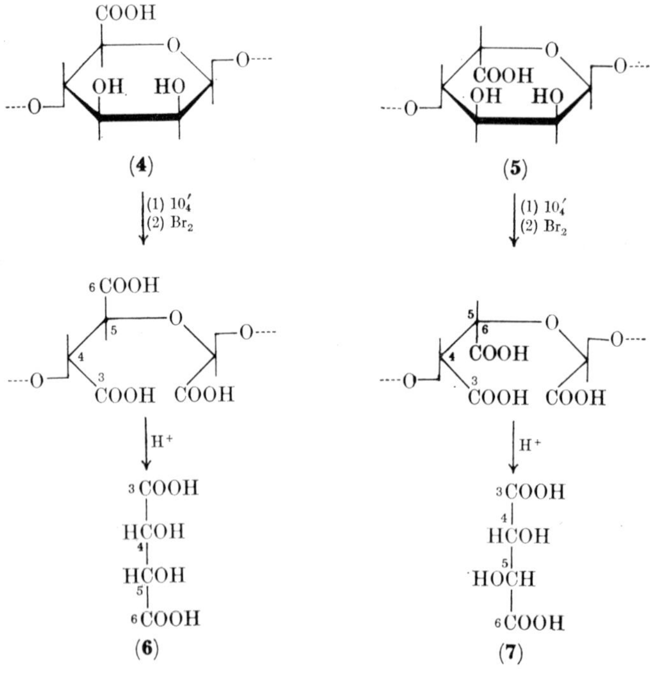

FIG. 2.

of the two substituted pentaric acids isolated after alkaline degradation of alginic acid (Whistler and BeMiller, 1960) supported this finding (cf. p. 43).

The 1,4-linkage of both uronic acids was confirmed by the characterization of 2,3-di-O-methyl-D-mannose [Fig. 3, (8)] and of 1,6-anhydro-2,3-di-O-methyl-β-L-gulopyranose [Fig. 3, (9)] as the crystalline phenylhydrazide and

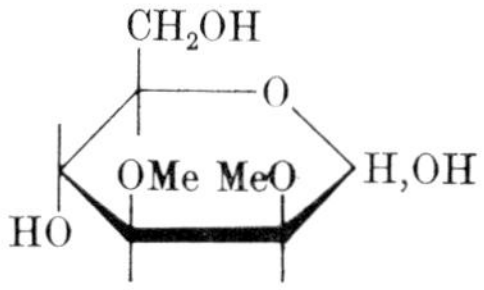

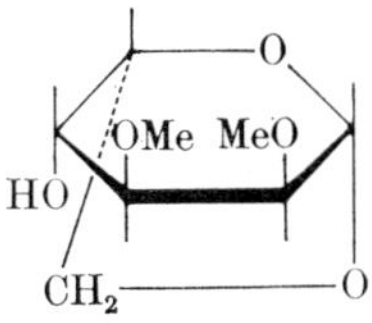

FIG. 3.

p-nitrobenzoate, respectively, from methylated alginic acid after hydrolysis, methyl ester methylglycoside formation and reduction with lithium aluminium hydride (Hirst and Rees, 1965). The failure to detect methylated guluronic acid derivatives in earlier studies is probably due to several causes. The 2,3-di-O-methyl erythraric acid [Fig. 4, (11)] considered to characterize 2,3-di-O-methylmannose (10) could equally well be derived from 2,3-di-O-methylgulose (12). Furthermore, 2,3-di-O-methylmannose and 2,3-di-O-

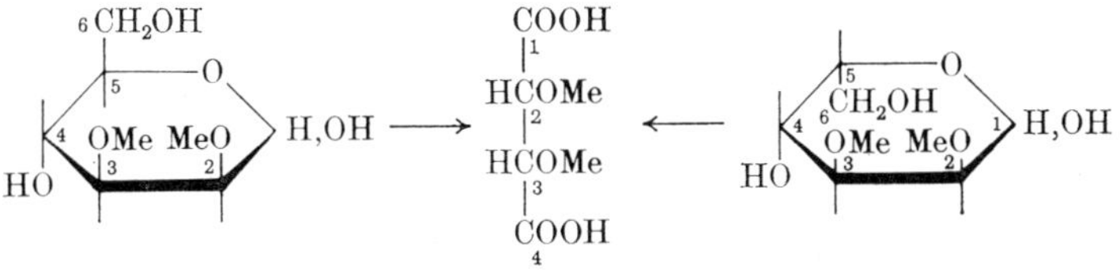

FIG. 4.

methylgulose are inseparable in the solvent systems used for chromatography by the earlier workers as also are the parent sugars, mannose and gulose obtained on demethylation. It is possible that the L-guluronic acid units are α-linked and that there are other minor linkages and some small amounts of other constituents present. Both glucuronic acid and xylose have been reported (Massoni and Duprez, 1960; Haug and Larsen, 1962), but it is very probable that these arise from contaminating polysaccharide (see p. 176).

It should be pointed out that in all the studies on alginic acid, theoretically yields of the two acids have never been obtained. Furthermore, instead of the expected reduction of one mole of periodate per anhydro unit, oxidation under controlled conditions stops after the reduction of about 0·6 mole per anhydro unit. Even if over-oxidation is allowed to proceed until the reduction reaches one mole of periodate for every uronic acid unit the polyaldehyde still contains uncleaved units of mannuronic acid and guluronic acid (Drummond *et al.*, 1962; Fujibayashi and Nisizawa, 1965).

It is not possible from these investigations to decide whether alginic acid is a heteropolymer comprising the two uronic acids or whether it consists of two different polymers, one made up entirely of mannuronic acid and the other solely of guluronic acid.

A. FRACTIONATION

Partial fractionation has been achieved by the addition of potassium chloride to solutions of sodium alginate (Haug, 1959, 1965). Material richer in mannuronic acid is precipitated. In contrast, addition of mixtures of manganous sulphate and potassium chloride give an insoluble fraction containing a higher proportion of guluronic acid than the original material (McDowell, 1958a). Although precipitation with calcium salts does not normally provide a suitable method of fractionation, the addition of calcium chloride to a 0·1% solution of sodium alginate in 0·08N to 0·5N-magnesium chloride gives a better fractionation than the methods used previously, and has the additional advantage of giving a more compact precipitate. Some fractionation can also be achieved by the use of magnesium chloride alone or in the presence of up to 6% ethanol (Haug and Smidsrod, 1965b). Nevertheless, repeated refractionation by any of these methods fails to yield material consisting solely of a single uronic acid. Only after degradation by heating with M-oxalic acid to give polymers of twenty-five to thirty units can a product approaching a single uronic acid polymer be separated.

B. PARTIAL HYDROLYSIS

The classical method for establishing the presence of both acids in a single molecule is the characterization of an aldobiouronic or triouronic acid, containing the two acids, from a partial acid hydrolysate of alginic acid. Unfortunately, like all polysaccharides with a high uronic acid content, alginic acid is very resistant to hydrolysis and conditions have to be so drastic that a considerable destruction of the uronic acid units takes place. Furthermore, the separation and characterization of a single pure oligouronic acid is also exceedingly difficult. Vincent (1960) separated a number of oligouronic acids containing both mannuronic and guluronic acids from a partial hydrolysate of alginic acid, but was unable to prove that any fraction constituted a single entity. A crystalline "so-called" dimannuronic acid has been isolated from a partial hydrolysate (Jayme and Kungstad, 1960), but the authors made no attempt to prove that was indeed its constitution.

In an attempt to settle this question, alginic acid was reduced to the neutral polymer before hydrolysis. This was achieved with diborane, but it was necessary first to convert the alginic acid into the dipropionate to render it soluble in the ether-type solvents used in the reduction (Hirst et al., 1964). At least 90% of the carboxyl groups in the alginic acid dipropionate were reduced to primary alcoholic groups and partial hydrolysis of this nearly neutral polymer led to the separation and characterization of crystalline mannosylgulose and β-1,4-mannobiose showing that at least some of the polymeric molecules contain both mannuronic and guluronic acid residues and also that adjacent mannuronic acid units are present. It is of interest to record that acid hydrolysis of the mannosylgulose gave rise to one mole of mannose and an equilibrium mixture (comprising one mole) of gulose and 1,6-anhydrogulose.

Recent studies by Haug et al. (1966) suggest that parts of the same molecule may be built up in different ways. By hydrolysing alginic acid with M-oxalic acid at 100° for ten-hour periods, they found that part of the material is readily hydrolysed to oligosaccharides which pass into solution leaving an insoluble residue, the amount of which is only slowly reduced by prolonged hydrolysis. Using alginic acid from L. digitata, about 28% of the alginate passed into solution in the first ten hours. When the insoluble residue was dissolved in alkali, reprecipitated as acid and hydrolysed for a further ten hours about 19% of it dissolved, and the same happened after a second dissolution and reprecipitation and a third ten-hour period of hydrolysis. The rate of formation of reducing end groups was the same during all three periods of hydrolysis, but while there was very rapid depolymerization in the first period suggesting random attack on glycosidic linkages, only a small reduction in the average degree of polymerization of the insoluble material took place in the second and third periods, in agreement with the removal of hydrolysable material mainly from chain ends.

The resistant material had a number average DP of twenty to thirty units and after dissolving in alkali it could be separated into two fractions by making the solution 0·1M in sodium chloride and adjusting its pH to 2·85. The insoluble portion was found to consist almost entirely of guluronic acid units and the soluble fraction of mannuronic acid units.

The hydrolysate from the first ten-hour period contained monomeric guluronic and mannuronic acids and diuronides with two different chromatographic mobilities which have not been characterized as yet. Chromatographic analysis of the soluble products obtained by further hydrolysis of the insoluble material remaining after two and three ten-hour periods gave only one diuronide spot. This moved at the same speed as the minor component in the first hydrolysate. The explanation is advanced that the diuronide from the later hydrolyses was a mixture of dimannuronic and diguluronic acids and that the major diuronide in the first hydrolysate contained a mannuronic and a guluronic residue.

It is therefore suggested that the molecule is built up of blocks of mannuronic acid and of guluronic acid units, twenty to thirty units long, separated by sections having a high proportion of alternating guluronic and mannuronic residues. The homogeneous sections are considered to be protected from random hydrolysis by their crystalline character.

Alginates from different sources vary in the proportion of material which is resistant to oxalic acid; for example, 60 to 70% for alginate from *L. digitata* and *L. hyperborea* and about 40% from *A. nodosum*. In this connection it is interesting that Myklestad and Haug (1966) have recently found that a water-soluble fraction can be separated from *A. nodosum*.

C. EQUIVALENT AND MOLECULAR WEIGHT

1. *Equivalent Weight*

The generally accepted structure of alginic acid as a polymer of anhydromannuronic acid and -guluronic acid residues should lead to an equivalent weight of 176 for the free acid. Figures close to this value have been obtained for alginic acid dried for 24 hr *in vacuo* over phosphorus pentoxide at 60°C (Chamberlain *et al.*, 1945), but samples dried to constant weight at 105° give values varying from 180 to 200.

Sodium alginate solutions obtained by titration of alginic acid with sodium hydroxide, evaporated to a film and then dried for 4 hr at 105°C consistently give figures close to 216, suggesting that one molecule of water is associated with each uronic unit.

2. *Molecular Weight*

The most convenient method of determining the molecular weight of a sample of an alginate is by measurement of the intrinsic viscosity of a dilute solution, but calibration by an absolute method is necessary (see p. 50). On account of the polyelectrolyte character of alginate solutions, it is necessary to include a simple electrolyte in such solutions used in extrapolating to zero concentration. A concentration of 0·1N-sodium chloride or carbonate is sufficient.

Donnan and Rose (1950) measured the intrinsic viscosities and osmotic pressures of a series of sodium alginate solutions and concluded that the degree of polymerization was given by the relation

$$DP = 58[\eta]$$

This figure assumes a unit weight of 216 for sodium alginate.

A critical examination of the various suggested relations between molecular weight and intrinsic viscosity was made by Cook and Smith (1954) using the ultracentrifuge and diffusion measurements. They concluded that a random coil model (Mandelkern-Flory) was the most suitable for sodium alginate, and on this basis obtained figures for the weight average molecular weight rather higher than those of Donnan and Rose for the number average molecular weight. This was considered to be a reasonable result as sodium alginate is polydisperse.

It is clear that there must be some slight uncertainty regarding average molecular weights of different sodium alginate samples determined by viscosity measurements, as the molecular weight distribution is not known. A DP of about 1000 for the most highly polymerized samples is probable.

D. X-RAY EXAMINATION

Alginic acid fibres prepared by extrusion of sodium alginate into a coagulating bath and stretched in the wet state show a sharp X-ray diagram indicating that the molecules are regularly orientated (Kringstad and Lunde, 1938; Astbury, 1945; Palmer and Hartzog, 1945; Tallis, 1950; Sterling, 1957; Warwicker, 1958). Although there is a general similarity to the cellulose diagram, Astbury found that the distance along the fibre axis was 8·7 Å compared with 10·3 Å for cellulose. At that time alginic acid was thought to be composed only of mannuronic acid residues, and the spacing was explained by the units being in the 1C conformation so that the glycosidic bonds were nearly at right angles to the plane of the ring. This conformation would bring the carboxyl groups into the axial position and measurements of birefringence of alginate gels (Sterling, 1957) suggest that this is unlikely. Furthermore, figures close to the Astbury spacing can be obtained with mannuronic acid in the C1 conformation, and the variations reported by different workers can be explained by small rotations about the glycosidic bonds.

At the same time it has been suggested (Frei and Preston, 1962) that the Astbury X-ray diagram is in fact that given by polyguluronic acid. The most stable conformation for L-guluronic acid is the 1C in which the carboxyl groups are in the equatorial position and this conformation gives shorter spacing in the manner originally suggested (Astbury, 1945; Percival and Percival, 1962).

In view of the work on the constitution of alginic acid, it is most unlikely that any of the samples examined by X-ray diffraction consisted of either pure polymannuronic or polyguluronic acid, and the interpretation of the results is by no means conclusive.

Calcium alginate does not give such sharp X-ray patterns as alginic acid (Tallis, 1950; Sterling, 1957), and is therefore less regular in structure. Certainly any intramolecular links (see p. 116) would pull the chains out of alignment.

It has also been found that to obtain the best diagrams, alginic acid must retain some moisture and intensive drying makes the pattern more diffuse (Astbury, 1945).

IV. ANALYTICAL METHODS

A. THE ALGINATE CONTENT OF A SEAWEED

In determining the alginate content of a seaweed it is usual to extract the alginate as the sodium salt (see Isolation, p. 100) and then determine its concentration in the derived solution by one of the following methods:

1. (a) If the concentration of alginate in the extract is more than about 0.1% and at least 100 mg of alginate is available, the most satisfactory method is by precipitating the alginate with an excess of calcium chloride, leaching free from calcium with hydrochloric acid, washing free from chloride and titrating the alginic acid with standard carbonate-free sodium hydroxide using phenolphthalein as indicator.

(b) Instead of direct titration of the alginic acid, which in some cases gives a viscous sodium alginate solution thus prolonging the time required for titration, calcium acetate solution can be added to the alginic acid and the liberated acetic acid titrated (Cameron et al., 1948). The direct titration method has the advantage of giving a sodium alginate solution which can be used for viscosity determinations after adding water and sodium salt to give the required alginate and electrolyte concentration.

2. For smaller amounts of alginic acid colorimetric methods are available:

(a) Carbazole in concentrated sulphuric acid was used by Percival and Ross (1948). A greatly enhanced absorption was produced by glucuronic and iduronic acids when sodium tetraborate is added to the concentrated sulphuric acid (Gregory, 1960), and this is also the case with mannuronic and guluronic acids (Elizabeth Percival, unpublished results). An automated procedure using this method has recently been developed by Balazs et al. (1964).

(b) Other workers have found that Bial's reagent (orcinol in hydrochloric acid) is more satisfactory (Brown and Hayes, 1952). Here again the absorption is increased by the addition of borate. Haug and Larsen (1962) have, however, pointed out that the values obtained by these methods depend to some extent on the proportions of the uronic acids present in the alginic acid.

3. On heating uronides with 19% hydrochloric acid, decomposition to carbon dioxide and furfural takes place, and the amount of carbon dioxide evolved can be used as a measure of the uronide present. Jensen et al. (1955) have adapted this method for determining the amount of alginic acid in seaweed. They consider that pre-treatment of the weed with sulphuric acid

removes interfering substances, but the results still depend on the rather doubtful assumption that alginic acid is the only uronide present.

Perlin (1952) has shown that carbon dioxide is liberated quantitatively when an alginate is heated without the addition of an acid.

B. THE ANALYSIS OF THE PROPORTION OF GULURONIC ACID AND MANNURONIC ACID IN SAMPLES OF ALGINATE

In addition to the fact that complete hydrolysis of alginic acid results in considerable degradation, particularly of guluronic acid residues, analysis is complicated by the fact that both uronic acids readily form lactones so that these as well as the free acids are present in a mixture after hydrolysis. In spite of these difficulties separation has been achieved by chromatography (Fischer and Dörfel, 1955), with pyridine-ethyl acetate-water (11:40:6, v/v) as eluant. The two acids and their lactones were eluted from the papers and determined with tetrazolium chloride. The standardization of the procedure is not easy and requires very careful technique for accurate results. Iono-phoretic separation can be obtained in 0·01M-sodium tetraborate solution (pH 9·2) containing 0·005M-calcium chloride in 2 hr with a current of 0·5 mA/cm (Haug and Larsen, 1961a). The mobilities of guluronic, mannuronic, galacturonic and glucuronic acids relative to that of glucose = 1 are 0·74, 0·88, 1·03 and 1·26, respectively.

A quantitative determination based on the use of columns of ion exchange resin has been developed by Haug and Larsen (1961b, 1962). These authors pre-treat the alginate with 80% sulphuric acid at 0° and then allow the mixture to stand at 20° for 18 hr. The acid is then diluted to 2N during cooling in ice and the mixture is heated in a sealed tube in boiling water for 5 hr. After neutralization with calcium carbonate any lactones are transformed to uronic acids by keeping the solution at pH 8 with alkali for 30 min and the acids are then separated on a Dowex 1 anion exchange resin column in the acetate form. The two acids are degraded at different rates during the hydrolysis. The authors assumed that losses incurred would be the same for both acids and concluded that by using this standard procedure and multiplying the ratio of mannuronic to guluronic acid by 0·66 (to allow for the greater degrada-tion of guluronic acid), a good estimate of the uronic acid composition of the alginate sample is obtained.

V. THE COMPOSITION OF ALGINATES FROM DIFFERENT SPECIES OF ALGAE

The D-mannuronic/L-guluronic (M/G) ratios of alginic acid from some of the more common species of brown algae, and from some representatives of

Table 1   Composition of alginates from different species of algae.

Species	M/G ratio		
	(a)		(b)
Ectocarpales			
Ectocarpus confervoides	0·4	(0·3)	—
Ectocarpus sp.	—		0·45
Sphacelariales			
Sphacelaria bipinnata	0·6	(0·4)	—
Dictyotales			
Dictyota dichotoma	0·6	(0·4)	1·05
Dictyopteris polypodioides	0·6	(0·4)	—
Chordariales			
Mesogloia vermiculata	—		0·25
Chordaria flagelliformis	—		0·90
Spermatochnus paradoxus	—		1·30
Dictyosiphonales			
Scytosiphon lomentaria	—		1·15
Dictyosiphon foeniculaceus	—		0·85
Desmarestiales			
Desmarestia aculeata	—		0·85
Laminariales			
Chorda filum	1·1	(0·8)	—
Laminaria digitata	3·1	(2·1)	1·45–1·6
Laminaria digitata (f)			1·2–1·85
Laminaria digitata (nf)			2·35
Laminaria hyperborea	1·6	(1·1)	
Laminaria hyperborea (s)	—		0·4–1·0
Laminaria hyperborea (f)	—		1·05–1·65
Laminaria hyperborea (nf)	—		1·90
Laminaria saccharina (f)	—		1·25–1·35
Alaria esculenta			1·20–1·70
Fucales			
Ascophyllum nodosum	2·6	(1·7)	1·40–2·25
Fucus vesiculosus	1·3	(0·9)	0·75–1·20
Fucus serratus	2·7	(1·8)	1·15
Pelvetia canaliculata	1·5	(1·0)	1·30–1·50
Himanthalia elongata	2·7	(1·8)	1·00–1·80
Halidrys siliquosa	1·1	(0·8)	0·75
Cystoseira barbata	0·7	(0·5)	—
Cystoseira abrotanifolia	1·9	(1·2)	—
Sargassum linifolium	0·8	(0·6)	—

(a) Fischer and Dörfel (1955). Figures in brackets are corrected by Haug's factor for the higher rate of destruction of guluronic acid.

(b) Haug (1964).

Whole plants were used for analysis except in the cases indicated as follows: (f) = fronds; (nf) = new fronds; (s) = stipes.

other orders of the Phaeophyceae are given in Table 1. The results are taken from the work of Fischer and Dörfel (1955) and Haug (1964). The figures in brackets are Fischer and Dörfel's multiplied by the factor 0·66, introduced by Haug and Larsen (1962) (see p. 109) to allow for the faster rate of breakdown of guluronic acid during hydrolysis.

There seems to be no correlation between the botanical classification and the alginic acid composition, and indeed the difference in composition of the new fronds and of the stipes of *L. hyperborea* comes near to covering the whole range of variation among the species examined. Agreement between the different authors' results is not very good in some cases but this is not surprising considering the variation found by Haug in different plants of the same species all collected round the Norwegian coast. The two results for *Dictyota dichotoma* are markedly different, but Fischer and Dörfel's sample was collected in the Bay of Naples while Haug's was of Norwegian origin.

Examination of different parts of the plant show variations related to some extent to the age of the tissues, as indicated by figures for mature and new fronds in Table 1, and for different parts of the stipe of *L. hyperborea* in Table 2.

The extremes of the M/G ratios obtained by fractionation (Haug, 1964; Haug and Smidsrod, 1965b) are given in Table 3. It will be seen that they are

Table 2   Composition of alginate from *L. hyperborea* stipes[b]

Part of stipe	M/G ratio: Lower part	Upper part
Medulla	0·70	0·92
Inner cortex	0·55	0·54
Outer cortex	0·44	0·43
Peripheral tissue	0·54	0·69
Weight average	0·46	0·57

[b] Haug (1964)

Table 3   M/G ratios for some alginate fractions.

Method	M/G ratios of fractions Soluble	Insoluble
From *L. digitata* (M/G = 1·6)		
Two fractionations with KCl[b]	1·05	2·3
Two fractionations with MnSO$_4$[b]	2·0	1·03
Fractionation with KCl followed by MnSO$_4$[b]	3·1	0·8
Two fractionations with 0·5N-MgCl$_2$+0·012N-CaCl$_2$[c]	1·05	2·9
From *L. hyperborea* (Outer cortex) (M/G = 0·44)		
One fractionation with MnSO$_4$[b]	1·03	0·43

[b] Haug (1964).
[c] Haug and Smidsrod (1965b).

not very different from ratios obtained for the whole of the alginic acid from some plants.

VI. ENZYMIC HYDROLYSIS

The little that is known about the synthesis and breakdown of alginic acid by the enzymes present in the algae is discussed in Chapter 1 (see p. 17).

No enzymes from higher animals and only a few from lower animals will degrade alginates (Oshima, 1931), and little activity has been found among fungi (Waksman and Allen, 1934).

Depolymerization of alginates by bacteria isolated from soil and from sea-water has been reported by several workers (Waksman et al., 1934; Kass et al., 1945) who observed a reduction in viscosity of the solution or lack of precipitation by calcium salts. Kooiman (1954) grew soil organisms with sodium alginate as the carbon source and obtained an enzyme preparation by precipitation of the toluene treated centrifugate with alcohol. Both this enzyme preparation and a culture filtrate reduced the viscosity of alginate solutions, and after two weeks incubation, paper chromatography showed the presence of a substance moving at the same speed as mannuronic acid as well as slower moving substances considered to be oligouronides.

The formation of unsaturated uronic acids in the enzymic breakdown of sodium alginate was observed by Tsujino and Saito (1961). One part of an aqueous extract of the homogenized liver of abalone (*Haliotis discus hannai*) was incubated with ten parts of a 1% solution of sodium alginate at 30° for 20 hr. Addition of sulphuric acid precipitated higher molecular weight materials, and from the filtrate an oligouronide, 10% by weight of the original sodium alginate, was obtained as colourless crystals by fractionation on a charcoal–Celite column. From its molecular weight, and an ultraviolet absorption band at 232 mμ which disappeared on bromination or reduction, it was concluded that the material was a diuronide with a 4–5 double bond. On hydrolysis, mannuronic acid and its lactone were detected by paper chromatography.

A more complete study of the products or enzymic degradation of sodium alginate was made by Preiss and Ashwell (1962a). A partially purified alginase isolated from a cell-free extract of a *Pseudomonas* which had been grown on alginic acid as the sole carbon source was found to degrade sodium alginate into a series of oligosaccharides containing a 4,5-unsaturated uronic acid on the non-reducing end of the oligosaccharide chain [see Fig. 5, (**13**)]. Due to the loss of the asymmetry at C-5 both D-mannuronic and L-guluronic acid yield the same unsaturated derivative (see Fig. 5). The unsaturated oligosaccharides are further degraded to yield finally 4-deoxy-L-erythro-5-hexoseulose uronic acid (4-deoxy-5-ketouronic acid) (**16**). [Cf. the bacterial degradation of mucopolysaccharides, Chondroitin sulphate A, B and C, and hyaluronic acid (Linker *et al.*, 1956, 1961) which results in the formation of 4-deoxy-L-threo-5-hexoseulose uronic acid, the 2-epimer of that derived from alginic acid.] However, using an enzyme which had been further purified, the oligosaccharides were not degraded further, so it is concluded that at least two enzymes were involved in breaking the alginate down to this monomer.

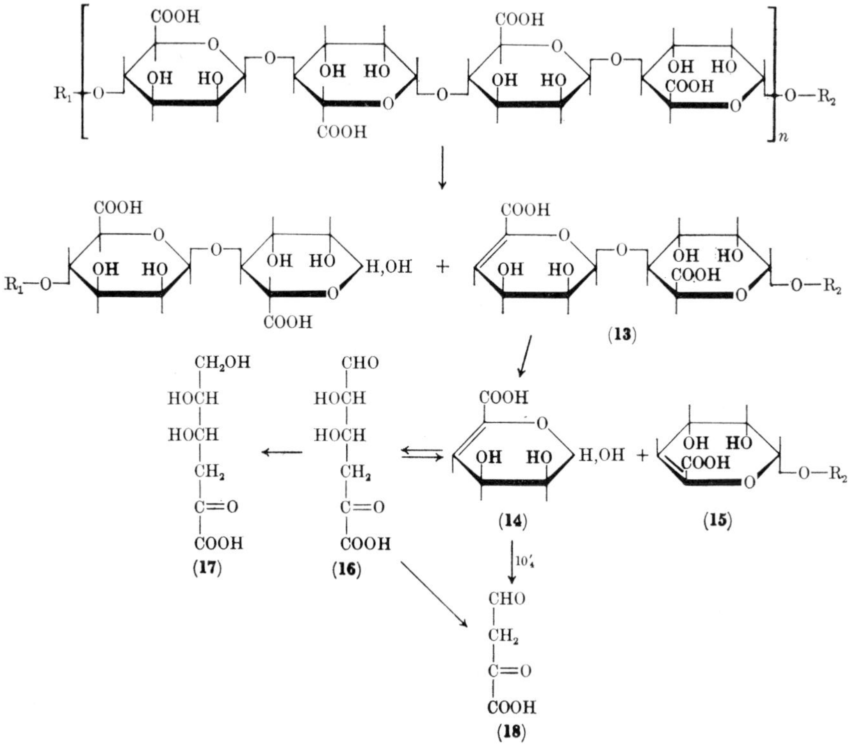

FIG. 5.

Oxidation of (**13 to 15**) with periodate gives β-formyl pyruvic acid (**18**) and treatment of this with thiobarbituric acid (Weissbach and Hurwitz, 1959) gives a colour reaction that can be used to measure the combined amounts of unsaturated residues in the degraded alginate (cf. alkaline degradation, p. 125) and of the free 4-deoxy-5-ketouronic acid. As little as 0·01 μmole of β-formyl pyruvate can be determined by this method.

The formation of considerable proportions of 4,5-unsaturated oligosaccharides which could be separated by paper chromatography in the reaction products indicate that an endo-enzyme, attacking the chain at random is present in the preparation. The action of different enzymes in the formation of unsaturated oligosaccharides and their breakdown to a 4-deoxy-5-ketouronic acid has been confirmed by other workers (Eimhjellen *et al.*, 1963).

A further enzyme isolated from alginate adapted *Pseudomonas*, a specific TPNH-linked dehydrogenase (Preiss and Ashwell, 1962b) reduced the alginase end-product (**16**) to 2-keto-3-deoxy-D-gluconate (**17**). This can be broken down by other enzymes in the presence of ATP to pyruvate and triose phosphate.

VII. PROPERTIES OF ALGINIC ACID AND ALGINATES

A. DISSOCIATION CONSTANT

The dissociation constant of alginic acid was studied by a number of workers before it was known that alginates from different sources contained different proportions of mannuronic and guluronic acid residues. More recent measurements (Haug, 1961a) made with these compositional variations in view gave the highest pK values for those alginates having the highest proportion of guluronic acid residues. This is in keeping with the pK values found for mannuronic acid (3·38) and guluronic acid (3·65). As with other polyacids, the pK value of alginic acid is dependent on the ionic strength of the solution. Measurements were made in 0·1N-sodium chloride giving values of 3·74 for alginic acid from *L. hyperborea* and 3·42 for the acid from *L. digitata*.

B. MOISTURE RELATIONS

Alginic acid and alginates retain moisture tenaciously, and complete drying is a very slow process (Chamberlain *et al.*, 1945). On exposure to humid air they pick up moisture, but again attainment of equilibrium is very slow, periods of six to nine months at room temperature being required, although changes in the moisture content are extremely slow in the later stages of exposure.

Samples of alginic acid and metal alginates exposed to the usual atmospheric conditions will contain between 10% and 25% moisture.

Dried insoluble alginates absorb water and swell when placed in aqueous solution, but the degree of swelling is extremely variable and the factors involved have not been extensively studied.

C. SOLUBILITY

Although alginates absorb moisture, there are strong forces holding the polymeric molecules together and it is only the highly ionized salts which dissolve in water. Solubility in non-aqueous liquids, or even in mixtures of them with water, is limited to the salts of some amines, and solution in non-polar solvents only takes place after considerable chemical modification of the alginate.

1. *Alginic Acid*

Alginic acid is insoluble in water, being precipitated by the addition of strong acids to alginate solutions. Recent work on products obtained by partial hydrolysis of alginic acids (Haug *et al.*, 1966) has shown that acidification to pH 2·85 of a solution of the degraded alginate in 0·1M-sodium chloride leads to precipitation of an alginic acid containing 80 to 90% guluronic acid while an acid composed of 80 to 90% mannuronic acid remains in solution. With alginates obtained by the normal extraction from seaweeds, the acid is precipitated over a pH range which depends on the species from which it was obtained as well as on the degree of polymerization and the ionic strength (Haug and Larsen, 1963). In general, no separation into fractions of different composition is observed, but it has been found possible to separate an acid soluble fraction from *A. nodosum* alginic acid (Myklestad and Haug, 1966). This material in solutions of low ionic strength is soluble over the whole pH range, but in the presence of alkali metal salts it is precipitated at about pH 2. Potassium chloride is the most effective precipitating agent, but even in N-potassium chloride it is soluble below pH 1·8 as well as above pH 3·1. It can be separated from the insoluble form by adjusting the pH to 1·4, and removing the precipitate by centrifuging. No chemical differences have been found between the soluble and insoluble forms, and both have been obtained over a similar range of molecular weights by starting with alginates of different intrinsic viscosities. The authors suggest that the difference in solubility is a result of a difference in the sequence of mannuronic and guluronic acid residues. The starting material was free from ascophyllan (see p. 176) which is also soluble in acids.

2. *Water-soluble Alginates*

Alginates of the alkali metals and ammonia and some low molecular weight organic bases are soluble in water. Unlike other divalent metals, magnesium also gives a water soluble alginate. None of these alginates has a clearly defined solubility limit and as the concentration of the alginate is increased, the mixtures with water change from viscous liquids to pastes or plastic solids, the concentration for the transition range depending on the degree of polymerization.

Solutions of alkali metal alginates are typical polyelectrolytes and the properties of their solutions are modified by changes in ionic strength. Small additions of electrolytes reduce the viscosity of dilute alginate solutions and, in some cases, larger amounts lead to precipitation. Alginates rich in man-nuronic acid are more readily precipitated than those of high guluronic acid content by potassium chloride and some fractionation takes place with suitable adjustment of conditions (Haug, 1965). On the other hand, alginates rich in guluronic acid are more readily precipitated by sodium chloride.

Most water-soluble alginates are precipitated by the addition of water miscible organic liquids such as alcohols and ketones. The amount required depends on the liquid used (the less polar the liquid, the less is needed) and on the base combined with the alginate. Ease of precipitation is in the order:

easier precipitation

←————————————————————

Mg   Na & K   NH$_4$ Amine salts

Alginates of some amines, for example diethylaminoethanol, require a very high proportion of one of the less polar solvents, such as acetone, for precipi-tation.

3. Insoluble Alginates

The alginates of most divalent and polyvalent metals are insoluble in water and organic solvents. They swell to some extent in water and, with few ex-ceptions, readily take part in reactions involving the metal ions. It is possible that the hydroxyl groups as well as the carboxyl groups play some part in holding the polymer chains together. Schweiger (1962b) found that partially acetylated alginates, even with an average of less than one acetyl group per

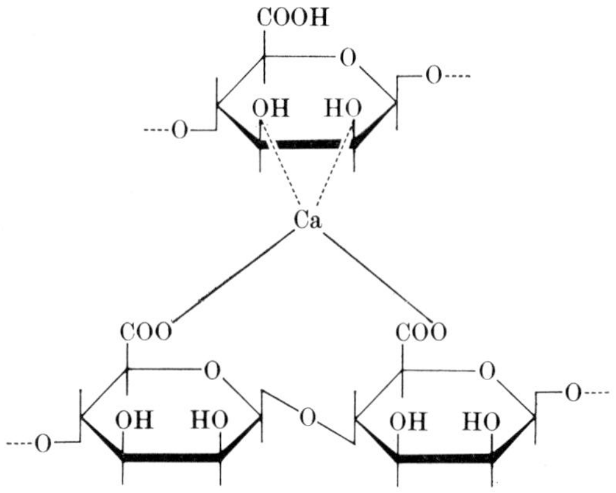

FIG. 6.

uronic unit, were not precipitated by calcium chloride. He, therefore, suggested that calcium forms ionic bonds between carboxyl groups on adjacent units in the same chain, and that for crosslinking between chains to take place, adjacent hydroxyl groups must be available for the formation of coordinate links with the calcium (Fig. 6). Examination of molecular models shows that it would be possible to form some links in this way, with both mannuronic and guluronic acid units, but that the chains would be very distorted. The X-ray pattern of calcium alginate is not very sharp, but the degree of crystallinity indicated by it would suggest that there is only a small proportion of crosslinks of this type. A different explanation for the lack of precipitation of acetylated alginic acid could be that the acetyl groups by their bulk keep the polymer chains from coming close enough for crosslinking.

D. VISCOSITY

An outstanding property of the water-soluble salts is the high viscosity of their solutions. If an alginate is extracted from a seaweed by methods which avoid degradation, a 1% solution can be expected to have a viscosity of between 500 and 3000 cps. The use of intrinsic viscosity measurements in determining molecular weight has already been mentioned and the viscosity at higher concentration leads to the use of alginates in a number of applications.

The relation between viscosity and concentration has been widely studied, but is complicated by the strong dependence of viscosity on rate of shear found in the more viscous solutions (McDowell, 1966). With the limited range studied by Donnan and Rose (1950), all sodium alginates could be represented by the same curve, with suitable adjustments to the concentration scale. Extension of this method to a wider range of viscosities and concentrations is unsatisfactory, even if comparisons are made at constant rates of shear or of shear stress. In the absence of divalent ions no evidence has been found for an effect due to the uronic acid proportions.

If a small proportion of the alkali metal in an alginate is replaced by a divalent metal, the resulting product gives solutions of higher viscosity than that of the corresponding alkali metal alginate. The magnitude of the effect depends on a number of factors. As would be expected, increasing proportions of divalent metal give higher viscosities, but the relative increase for a given proportion is much greater at high than at low concentrations of alginate in water, and also increases with the molecular weight of the alginate. A high proportion of mannuronic acid units in the alginate also favours viscosity increases with divalent ions. The viscosity of these solutions of mixed metal alginates is highly dependent on the rate of shear.

5+

E. ALGINATE GELS

Insoluble alginates in a freshly precipitated form are highly hydrated and over a considerable range of concentration they can be obtained as uniform gels. They are generally prepared by the slow release of precipitating ions in an alginate solution. An example is the use of δ-D-gluconolactone and calcium diphosphate with a sodium alginate solution: as the lactone hydrolyses to give gluconic acid, calcium ions are made available from the phosphate and react with the alginate throughout the solution to give uniform gel structure. As would be expected, the gel strength increases with the concentration and degree of polymerization of the alginate, but it is also dependent on the algal species from which the alginate was obtained, presumably as a consequence of the different proportions of mannuronic and guluronic residues present. The highest gel strengths are obtainable with alginate from *L. hyperborea*, which, of all the algae studied for this purpose, has the highest proportion of guluronic acid residues.

If an alginate gel is formed by the diffusion of a divalent metal ion into a sodium alginate solution, the gel formed is birefringent (Thiele and Andersen, 1955) indicating that there is orientation of the polymer chains during the precipitation process. The gel formed has a smaller volume than the original alginate solution, and the greater the shrinkage, the greater is the birefringence.

Other conditions being equal, the shrinkage and birefringence are dependent on the metal used and can be arranged in the ionotropic series:

$$\text{Pb} \quad \text{Cu} \quad \text{Cd} \quad \text{Ba} \quad (\text{Sr Ca}) \quad (\text{Co Ni}) \quad \text{Zn} \quad \text{Mn}$$
$$\xleftarrow{\hspace{6cm}}$$
Increasing shrinkage and birefringence

It should be pointed out that this series is based on results from *L. digitata* alginate and is in the same order as the selectivity coefficients for this alginate (see p. 119).

Orientated calcium alginate gels can be obtained by allowing calcium chloride to diffuse through a membrane into a flowing stream of sodium alginate solution. Their birefringence and X-ray patterns have been studied (Sterling, 1957).

F. BASE EXCHANGE REACTIONS

Insoluble alginates take part in base exchange reactions, and for the reaction:

$$\underset{\text{gel}}{\text{Me}(\text{Alg})_2} + \underset{\text{liquid}}{2\text{Na}^+} \rightleftharpoons \underset{\text{gel}}{2\text{NaAlg}} + \underset{\text{liquid}}{\text{Me}^{++}}$$

where Me^{++} is a divalent cation.

An equilibrium

$$\frac{[\text{Me gel}][\text{Na}^+\text{liq.}]^2}{[\text{Na gel}]^2[\text{Me}^{++}\text{liq.}]} = K \text{ is set up.}$$

K is termed the selectivity coefficient when the concentrations in the gel are expressed as equivalent fractions and those in the liquid as normalities (Haug, 1961b; Haug and Smidsrod, 1965a).

The earlier studies on ion exchange (Mongar and Wassermann, 1952) were made before the variation in composition of alginates from different sources was known, but more recent work by the Norwegian group (Haug and Smidsrod, 1965a) has shown that there are significant differences in selectivity coefficients depending on the mannuronic/guluronic ratio. There is a wide difference also in the selectivity coefficients of different metals as shown in the Table 4.

Table 4 Selectivity coefficients.

Metal ions	L. digitata	L. hyperborea stipe
$Cu^{++} Na^+$	230	340
$Ba^{++} Na^+$	21	52
$Ca^{++} Na^+$	7·5	20
$Co^{++} Na^+$	3·5	4

The selectivity coefficients of most of the metals studied fall in the same numerical order as the ionotropic series of Thiele (p. 118). On the other hand, strontium has a high affinity for alginates with a major proportion of guluronic acid, so that it stands before cadmium in the series with these alginates. There is a close correlation between the selectivity coefficient of the Sr–Cd ion exchange and the composition of the alginate (Haug, 1961b).

G. PRECIPITATION REACTIONS

When an alginate is precipitated by adding the salt of a divalent metal to a solution of a soluble alginate, the base exchange equilibria have to be considered as well as the proportion of the metal needed to render the mixed alginate insoluble. In most instances, the order of effectiveness in making alginates insoluble is the same as that of the ionotropic series and of the selectivity coefficients, but barium is more effective than copper or lead in bringing about precipitation (Haug and Smidsrod, 1965a).

The sodium ions derived from the sodium alginate and any sodium salt added have two different effects. The increase in ionic strength leads to easier precipitation of the alginate, but the base exchange reaction prevents the whole of the divalent metal being combined with the alginate. The result is that with metals such as copper and lead, having a high selectivity coefficient, less of the metal is required to bring about precipitation when sodium salts are added, but more of such metals as nickel and cobalt, having a low selectivity coefficient, is required. With some metals, having intermediate values of selectivity coefficient, the effect depends on the concentration of the alginate, added sodium salts promoting precipitation at higher alginate concentration and hindering it in dilute solution (McDowell, 1958b).

VIII. DERIVATIVES

Apart from derivatives of alginic acid and its breakdown products which have been made in structural investigations, a number of compounds have been prepared in order to investigate their properties or in the search for useful products. Some are manufactured commercially.

A. SALTS OF ORGANIC BASES

Of the vast number of possible salts in which alginic acid may be combined with organic bases only a few have been described. They have been prepared mainly with the object of giving greater solubility in organic solvents. Triethanolamine alginate is an article of commerce, and is soluble in about 75% aqueous ethyl alcohol. Tributylamine, phenyltrimethylammonium and benzyltrimethylammonium alginates are soluble in absolute ethanol (McNeely, 1954). The alginate of "Primene 81R" (a mixture of amines with fourteen to twenty carbon atoms, having a primary amine group linked to a tertiary carbon atom) is soluble in polar organic solvents, but is precipitated on addition of water (Boyle, 1960).

Most of the quaternary ammonium compounds which have a long hydrocarbon chain, for example, cetyltrimethylammonium bromide, give a precipitate when added to an alginate solution (Scott, 1965).

B. ESTERS

1. Carboxyl Group Esterified

The methyl ester of alginic acid has been prepared by treatment of alginic acid with diazomethane and with methanolic hydrogen chloride (Lucas and Stewart, 1940; Jansen and Jang, 1946), and by the reaction of dimethyl sulphate with sodium alginate suspended in a non-aqueous liquid (McNeely and O'Connell, 1958).

1,2-Alkylene oxides react with alginic acid under mild conditions so that esters are formed with little degradation (Steiner and McNeely, 1951). By far the most important is the propylene glycol ester which is made with the readily available propylene oxide and is generally accepted as a safe food additive (which is not the case with esters of ethylene glycol). Unlike salts of alginic acid, it is not precipitated by acids, and will give viscous solutions in acid conditions.

2. Hydroxyl Groups Esterified

Acetyl alginic acid has been prepared by a number of workers, but in most cases, the reaction has involved considerable degradation. An improved method which minimizes degradation is reaction with a mixture of acetic acid and acetic anhydride using perchloric acid as catalyst (Schweiger, 1962a). Products with a degree of acetylation of up to 1·85 were obtained at a temperature of 45°.

Alginic acid sulphates have been made by reaction with chlorsulphonic acid using pyridine as a catalyst (Snyder, 1950). The emphasis with this derivative has been on the preparation of products with a low degree of polymerization which are suitable for use as blood anticoagulants, although highly polymerized products have been made (Schweiger, 1966).

The preparation of many other esters of alginic acid has been reported but none has found commercial application.

C. ETHERS

Methyl ethers have been prepared in work on the constitution of alginic acid (see p. 101). The sodium salt of carboxymethyl alginic acid can be made by interaction of sodium alginate with chloracetic acid in the presence of sodium hydroxide (McNeely and O'Connell, 1959), as in the preparation of carboxymethyl cellulose.

D. AMIDE

While the action of aqueous ammonia on glycol esters of alginic acid is saponification of the ester and degradation of the alginate, treatment with anhydrous ammonia under pressure gives an amide (Henkel, 1957).

IX. DEGRADATION

Alginic acid is depolymerized by several different mechanisms depending on conditions, and the rate of degradation increases very considerably when the pH is taken above 10 or below 5.

A β-elimination reaction with the formation of $\alpha\beta$-unsaturated uronic acid derivatives (cf., enzymic hydrolysis p. 112 and Fig. 5) is the principal degradation process at high pH values accounting for about 80% of the chain breakage at pH 10·8, but less than 10% at pH 5 (Haug et al., 1963). At still higher pH values, where degradation is more rapid, the proportion of bonds broken by β-elimination appeared to be only about 50%, probably due to further breakdown of the unsaturated compounds in strongly alkaline conditions. The extent of the β-elimination reaction can be measured by periodate oxidation of the resulting polymer, hydrolysis and estimation of the derived formyl pyruvic acid with thiobarbituric acid (Weissbach and Hurwitz, 1959).

The formyl pyruvic acid appears to have been derived solely from oligosaccharides produced by a random attack on the macromolecule as no dialysable material giving the thiobarbituric acid reaction was obtained. (Degradation to $[\eta] > 3$.)

As would be expected from analogy with other β-elimination reactions, saponification of esters of alginic acid by alkali leads to considerable degradation with formation of the unsaturated acids (R. H. McDowell, unpublished work).

The rate of proton catalysed hydrolysis of alginic acid increases as the pH is reduced, and is the dominant mechanism at pH values below 5 (Haug et al., 1963).

It is noteworthy that in the pH range 1 to 4 alginic acid, in common with other uronides, is hydrolysed more rapidly than neutral polysaccharides whereas, in more acid conditions, the reverse is the case (Smidsrod et al., 1966) (see also p. 32).

Alginates are also degraded by an oxidative-reductive depolymerization similar to that found for a number of other polymers (Smidsrod et al., 1963a, 1963b). Although the degradation takes place as a result of reducing substances, such as phenolic compounds present in some algae, in the alginate solution, it appears that the formation of hydrogen peroxide plays an essential part in the reaction. In the presence of added hydrogen peroxide, ascorbic acid and ferrous ions are particularly effective in increasing the rate of degradation. It is considered that the effective agent in bringing about the reaction is the hydroxyl radicle (Smidsrod et al., 1965).

X. APPLICATIONS

The main properties of alginates on which their widespread uses are based are:

(a) The formation of viscous solutions at relatively low concentrations;
(b) Their behaviour as polyelectrolytes in solution;
(c) The formation of gels by chemical reaction;

(d) The formation of films on surfaces;

(e) The formation of films and fibres;

(f) Base exchange properties.

Alginic acid and its salts and the propylene gylcol ester are recognized as safe ingredients in food in the regulations of many countries including the United States and Great Britain. The principal food uses are those in which the alginate functions as a stabilizer, probably a result of a combination of polyelectrolyte character and raising viscosity. Among foods widely stabilized with alginates either alone or with other colloids are ice cream, water ices, artificial creams and many semi-solid products where phase separation is to be prevented. Edible jellies of various types are made by reaction with calcium salts.

Gel formation is also used in dental impression materials, and some semi-gels and viscous solutions in pharmaceutical and cosmetic products.

The most important use of sodium alginate as a viscosity increasing substance is in thickenings for textile printing pastes. It is the most satisfactory thickener for the cellulose-reactive dyes, first introduced in 1956.

The film forming property is used in the surface sizing of paper and it controls viscosity and water penetration.

Although alginates are not used as textile fibres in their own right, calcium alginate yarn has an important use in textile applications where a temporary thread is required. It can be dissolved away by the action of soap or soda solution.

These and many other uses have been described in detail in a number of publications (Maass, 1959; McDowell, 1960). Other applications have come into use or been suggested more recently. Small quantities of alginate of the order of 0·1 ppm have a marked effect in improving the flocculation in the purification of water with the aid of aluminium or iron salts (Hall, 1964).

Sodium alginate has been found particularly effective in selectively preventing absorption of strontium from the gut (Skoryna et al., 1964; Paul et al., 1964), and should therefore be valuable in reducing the ill-effects of the accidental ingestion of Sr^{90}. This has been confirmed by experiments on human subjects in which the short-lived radioisotope Sr^{85} was used (Hesp and Ramsbottom, 1965).

Propylene glycol alginate will react with many amino-compounds in alkaline conditions, and the reaction with gelatine has useful technical applications, particularly in photography (Agfa, 1964). If a film containing gelatine and propylene glycol alginate is treated with alkali it will not melt in contact with hot water, thus allowing processing in warm conditions. Although the mechanism of the reaction has not been fully investigated, it is probable that combination of the alginate with animo groups takes place as the ester group is removed by hydrolysis.

REFERENCES

Agfa, A. G. (1964). B. Pat. 962,483.

Astbury, W. T. (1945). *Nature Lond.* **155**, 667.

Balazs, E. A., Berntsen, K. O., Karossa, J., and Swan, D. A. (1964). *Analyt. Biochem.* **12**, 547.

Black, W. A. P. (1950). *J. Mar. biol. Ass. U.K.* **29**, 45.

Black, W. A. P. (1953). Chem. Soc. Ann. Reps., p. 332 and ref. cited therein.

Boyle, J. L. (1960). B. Pat. 835,009.

Brown, E. G., and Hayes, T. J. (1952). *Analyst, Lond.* **77**, 445.

Cameron, M. C., Ross, A. G., and Percival, E. G. V. (1948). *J. Soc. chem. Ind., Lond.* **67**, 161.

Chamberlain, N. H., Johnson, A., and Speakman, B. (1945). *J. Soc. Dyers Colour.* **61**, 13.

Chanda, S. K., Hirst, E. L., Percival, E. G. V., and Ross, A. G. (1952). *J. chem. Soc.* p. 1833.

Cook, W. H. and Smith, D. B. (1954). *Can. J. Biochem. Physiol.* **32**, 227.

Donnan, F. G., and Rose, R. C. (1950). *Can. J. Res.* B**28**, 105.

Drummond, D. W., Hirst, E. L., and Percival, Elizabeth (1958). *Chemy. Ind.* p. 1088.

Drummond, D. W., Hirst, E. L., and Percival, Elizabeth (1962), *J. chem. Soc.* p. 1208.

Eimhjellen, K. E., Rosness, P. A., and Hegge, E. (1963). *Acta chem. scand.* **17**, 901.

Fischer, F. G., and Dörfel, H. (1955). *Hoppe-Seyler's Z. physiol. Chem.* **301**, 224; **302**, 186.

Frei, E., and Preston, R. D. (1962). *Nature, Lond.* **196**, 130.

Fujibayashi, S., and Nisizawa, K. (1965). *J. Biochem. Japan.* **37**, 690.

Gregory, J. D. (1960). *Archs Biochem. Biophys.* **89**, 157.

Hall, R. I. (1964). *Proc. Soc. Wat. Treat. Exam.* **13**, 114.

Haug, A. (1959). *Acta chem. scand.* **13**, 601.

Haug, A. (1961a). *Acta chem. scand.* **15**, 950.

Haug, A. (1961b). *Acta chem. scand.* **15**, 1794.

Haug, A. (1964). "Composition and Properties of Alginates", Report No. 30, Norwegian Inst. of Seaweed Research, Trondheim.

Haug, A. (1965). *In* "Methods of Carbohydrate Chemistry" (R. C. Whistler, ed.), Vol. V, p. 69, Academic Press, New York and London.

Haug, A., and Larsen, B. (1961a). *Acta chem. scand.* **15**, 1395.

Haug, A., and Larsen, B. (1961b). *Acta chem. scand.* **15**, 1397.

Haug, A., and Larsen, B. (1962). *Acta chem. scand.* **16**, 1908.

Haug, A., and Larsen, B. (1963). *Acta chem. scand.* **17**, 1653.

Haug, A., and Smidsrod, O. (1965a). *Acta chem. scand.* **19**, 341.

Haug, A., and Smidsrod, O. (1965b). *Acta chem. scand.* **19**, 1221.

Haug, A., Larsen, B., and Smidsrod, O. (1963). *Acta chem. scand.* **17**, 1466.

Haug, A., Larsen, B., and Smidsrod, O. (1966). *Acta chem. scand.* **20**, 183.

Henkel *et cie* GMBH (1957). B. Pat. 768,309.

Hesp, R., and Ramsbottom, B. (1965). *Nature, Lond.* **208**, 1341.

Hirst, Sir Edmund, and Rees, D. A. (1965). *J. chem. Soc.* p. 1182.

Hirst, E. L., Jones, J. K. N., and Jones, W. O. (1939). *J. chem. Soc.* p. 1880.

Hirst, E. L., Percival, Elizabeth, and Wold, J. K. (1964). *J. chem. Soc.* p. 1493.

Jansen, E. F., and Jang, R. (1946). *J. Am. chem. Soc.* **68**, 1475.

Jensen, A., Sunde, I., and Haug, A. (1955). "The Quantitative Determination of Alginic Acid", Rep. No. 12, Norwegian Inst. Seaweed Research, Trondheim.

Jayme, G., and Kringstad, K. (1960). *Ber. dt. chem. Ges.* **93**, 2263.

Kass, E., Lid. I., and Molland, J. (1945). *Avh. norske Vidensk Akad. Oslo.* Matematisk naturwidenskapelig Klasse No. 11, p. 1.

Kooiman, P. (1954). *Biochim. biophys. Acta* **13**, 338.

Kringstad, H., and Lunde, G. (1938). *Kolloidzeitschrift* **83**, 202.

Linker, A., Meyer, K., and Hoffman, P. (1956). *J. biol. Chem.* **219**, 13.

Linker, A., Hoffman, P., Meyer, K., Sampson, P., and Korn, E. D. (1961). *J. biol. Chem.* **235**, 3061.

Lucas, H. J., and Stewart, W. T. (1940a). *J. Am. chem. Soc.* **62**, 1070.

Lucas, H. J., and Stewart, W. T. (1940b). *J. Am. chem. Soc.* **62**, 1792.

Maass, H. (1959). "Alginsaure und Alginate", Strassenbau Chemie & Technik Verlags-gesellschaft m.b.H., Heidelberg.

McDowell, R. H. (1958a). *Chemy. Ind.* p. 1401.

McDowell, R. H. (1958b). *Abstracts 3rd Int. Seaweed Symp.*, Galway.

McDowell, R. H. (1960). *Rev. pure appl. Chem.* **10**, 1.

McDowell, R. H. (1966). *In* "The Chemistry and Rheology of Water Soluble Gums", S.C.I. Monograph, No. 24, p. 19, Soc. Chem. Ind. Lond.

McNeely, W. H. (1954). U.S.P. 2,688,598.

McNeely, W. H., and O'Connell, J. J. (1958). U.S.P. 2,860,130.

McNeely, W. H., and O'Connell, J. J. (1959). U.S.P. 2,902,479.

Massoni, R., and Duprez, G. (1960). *Chimie Ind.* **83**, 79.

Mongar, I. L., and Wassermann, A. (1952). *J. chem. Soc.* p. 492.

Myklestad, S., and Haug, A. (1966). *Proc. 5th int. Seaweed Symp.* (1965), Halifax, Nova Scotia (E. G. Young and J. L. McLachlan, eds), p. 297, Pergamon Press, Oxford.

Nelson, W. L., and Cretcher, L. H. (1929). *J. Am. chem. Soc.* **51**, 1914.

Nelson, W. L., and Cretcher, L. H. (1930). *J. Am. chem. Soc.* **52**, 2130.

Oshima, K. (1931). *J. agric. chem. Soc. Japan* **7**, 328.

Palmer, K. J., and Hartzog, M. B. (1945). *J. Am. chem. Soc.* **67**, 1865.

Paul, T. M., Waldron Edward, D., and Skoryna, S. C. (1964). *Can. med. Ass. J.* **91**, 553.

Percival, E. G. V., and Percival, Elizabeth (1962). "Structural Carbohydrate Chemistry", p. 297, J. Garnet Miller, London.

Percival, E. G. V., and Ross, A. G. (1948). *J. Soc. Chem. Ind.* **67**, 420.

Perlin, A. S. (1952). *Can. J. Chem.* **30**, 278.

Preiss, J., and Ashwell, G. (1962a). *J. biol. Chem.* **237**, 309.

Preiss, J., and Ashwell, G. (1962b). *J. biol. Chem.* **237**, 317.

Rose, R. C. (1951). *Can. J. Technol.* **29**, 19.

Schweiger, R. G. (1962a). *J. org. Chem.* **27**, 1786.

Schweiger, R. G. (1962b). *J. org. Chem.* **27**, 1789.

Schweiger, R. G. (1966). *Chemy Ind.* p. 900.

Scott, J. E. (1965). *In* "Methods of Carbohydrate Chem." (R. C. Whistler, ed.), Vol. V, p. 38, Academic Press, New York and London.

Skoryna, S. C., Paul, T. M., and Waldron Edward, D. (1964). *Can. med. Ass. J.* **91**, 285.

Smidsrod, O., Haug, A., and Larsen, B. (1963a). *Acta chem. scand.* **17**, 1473.

Smidsrod, O., Haug, A., and Larsen, B. (1963b). *Acta chem. scand.* **17**, 2628.

Smidsrod, O., Haug, A., and Larsen, B. (1965). *Acta chem. scand.* **19**, 143.

Smidsrod, O., Haug, A., and Larsen, B. (1966). *Acta chem. scand.* **20**, 1026.

Snyder, E. G. (1950). B. Pat. 676,564.

Stanford, E. C. C. (1883). *Chem. News, Lond.* **96**, 254.

Steiner, A. B., and McNeely, W. H. (1951). *Ind. Engng. Chem.* **43**, 2073.

Steiner, A. B., and McNeely, W. H. (1954). *In* Am. Chem. Soc. Advances in Chem. Series No. 11, 72 and ref. cited therein.

Sterling, C. (1957). *Biochim. Biophys. Acta* **26**, 186.

Tallis, E. (1950). *J. Text Inst.* **41**, T151.

5*

Thiele, H., and Andersen, G. (1955). *Kolloidzeitschrift.* **140**, 76.
Tsujino, I., and Saito, T. (1961). *Nature, Lond.* **192**, 970.
Vincent, D. L. (1960). *Chemy. Ind.* p. 1109.
Waksman, S. A., and Allen, M. C. (1934). *J. Am. chem. Soc.* **56**, 2701.
Waksman, S. A., Carey, C. L., and Allen, M. C. (1934). *J. Bact.* **28**, 213.
Warwicker, J. O. (1958). *Shirley Inst. Mem.* **31**, 41.
Wassermann, A. (1948). *Ann. Bot.* **12**, 137.
Wassermann, A. (1949). *Ann. Bot.* **13**, 79.
Weissbach, A., and Hurwitz, J. (1959). *J. biol. Chem.* **234**, 705.
Whistler, R. L., and Schweiger, R. (1958). *J. Am. chem. Soc.* **80**, 5701.
Whistler, R. L., and BeMiller, J. N. (1960). *J. Am. chem. Soc.* **82**, 457.

CHAPTER 6

Sulphated Polysaccharides Containing Neutral Sugars: I Galactans of the Rhodophyceae

Polysaccharides with sulphate hemi-ester groups attached to sugar units are found in the form of fucoidan in the Phaeophyceae, as galactans in the Rhodophyceae, and as arabinogalactans with lesser amounts of other sugars in the Chlorophyceae. Because of the variety in the fine structure and the extent of the investigations on the galactans, they will be discussed in this chapter and fucoidan and the arabinolgalactans in the following chapter.

Recent studies on extracts from the brown seaweed *Ascophyllan nodosum* have revealed sulphated fucose-containing polysaccharides which also comprise considerable proportions of xylose and glucuronic acid. The Ulvaceae family of the Chlorophyceae synthesize similar material except that rhamnose replaces fucose, and *Phaeodactylum tricornutum*, the only member of the Bacillariophyceae to be investigated in detail chemically, also synthesizes a sulphated polysaccharide containing glucuronic acid. For convenience, it has been decided to discuss all these glucuronic acid containing polysaccharides in Chapter 8, although examination of the fucoidan in different genera of the Phaeophyceae may reveal that the fucose containing polymers so far investigated and considered here as two separate groups represent only the extreme ends of a whole range of similar polysaccharides which differ from one another in the fine details of structure.

I. INTRODUCTION

The characteristic polysaccharides of the red algae are mucilages which contain varying proportions of D- and L-galactose, 3,6-anhydro-D- and L-galactose, monomethylgalactoses and ester sulphate (Peat and Turvey, 1965; Rees, 1965). They appear to occur both in the cell wall and inter-cellularly. For convenience, they will be discussed under three main groups typified by agar-, porphyran- and carrageenan-type polysaccharides, although, as will appear later, a whole spectrum of polymers based on the above units have now been isolated and characterized from the different genera of

Rhodophyceae. There seems to be a family of galactans comprising alternate $1 \rightarrow 3$ and $1 \rightarrow 4$ linked galactose units which individually differ in their finer details of structure, and, as suggested by D. A. Rees, "The particular difference is appropriate for a particular species growing in a particular environment".

From the point of view of their properties and use, the classification of the red algae as sources of extractives is not completely in accordance with what is known of their chemical composition. Several classification schemes have been put forward (Stoloff and Silva, 1957; Hoppe, 1966), and the algae are grouped as giving agar, carrageen or gelan type extractives. The agar type may be subdivided as being true agars which conform to the specification of the U.S.A. Pharmacopoeia (see p. 152) and agaroids which are similar but fail to conform in some respect to this specification for physical properties (Selby and Selby, 1959). Carrageenans give gels in the presence of certain salts, react with protein and are precipitated by methylene blue, while gelans gel similarly but do not show the same precipitation reactions. There is general agreement on the species which yield the true agars, but chemically the position is complicated by the fractionation of agar into agarose and agaropectin. The placing of some other species among the groups has varied in the different schemes; some of the species concerned yield extractives which would be considered chemically among the agars, while others yield varying proportions of κ- and λ-carrageenan. None of the species used commercially for the preparation of extractives has yet been shown to contain polysaccharides of the porphyran-type.

Agar is synthesized by species of *Gelidium, Gracilaria, Acanthopeltis, Ahnfeltia, Ceramium, Campylaephora, Phyllophora* and *Pterocladia* spp., which may be classed together as agarophytes. The extract from the seaweeds has been an article of commerce in Japan since the seventeenth century and was introduced into Europe about the middle of the last century. Japan remains the largest producer of agar, making quantities of the order of two thousand tons annually. Other important producers are Spain, Korea, Russia and the United States of America, but their combined output is less than that of Japan (Ohmi, 1958). The traditional Japanese agar is prepared from a mixture of red seaweeds with *Gelidium amansii* being used in the largest quantity. However, *Gracilaria verrucosa* (*confervoides*), besides being a component of seaweed mixtures, is used as the sole source by several modern factories (Ohmi, 1958). It was used in the United States and some other countries during the Second World War when Japanese agar was unavailable, but at the present time there is no information on its use outside Japan. *Gelidium cartilagineum* and other *Gelidium* species are used by the American industry, while *Phyllophora nervosa* which grows in the Black Sea is the main species used in the U.S.S.R.

The important extractive made in Denmark from *Furcellaria fastigiata* is frequently referred to as agar, but chemical investigation shows that it is more closely allied to carrageenan (see p. 137). Production has steadily increased since the Second World War and was over 800 tons in 1960 (Lund and Bjerre-Petersen, 1964).

Species of *Chondrus* and *Gigartina* are used for the preparation of carrageenan. The largest quantities are made in the United States of America, approximately 1000 tons being produced there in 1956 (Stoloff, 1959). Other important producing countries are Denmark, France and Norway.

Somewhat similar extractives are obtained from *Eucheuma* and *Hypnea* species, but the quantities produced are relatively small. Another red seaweed of commercial importance in China and Japan is *Gloiopeltis furcata* which disperses completely in hot water to give a viscous non-gelling solution ("funori") which is used as an adhesive and for textile and paper sizing.

Several species of *Porphyra* are used as food in many parts of the world; *Porphyra perforata* (California laver) is commonly harvested from Alaska to Southern California. In Japan the demand for this type of food is so great that *Porphyra tenera* is extensively cultivated; the plants, after harvesting, are chopped, washed and the resulting dark-brown membranous sheets dried. The resulting product ("nori") is eaten as a hot delicacy and used in flavouring. In Wales, *Porphyra laciniata*, which is abundant on the Welsh coast, fulfils the same purpose under the name "laver bread". The extract from this group of seaweeds has not been made commercially, probably because the plants themselves have a ready market.

II. CONSTITUTION AND STRUCTURE OF THE DIFFERENT GALACTANS

A. AGAR

Agar is extracted from the agarophytes by boiling them in water. On cooling, the filtered solution sets to a gel, and this can be purified by freezing and thawing. Much of the water, containing soluble impurities, is removed from the gel by this procedure.

This material comprises two polysaccharides, agarose and agaropectin. The original separation was achieved after acetylation of the agar with acetic acid and pyridine (Araki, 1937a; Hjerten, 1961). The product was fractionated with chloroform into soluble agarose acetate (70%) and insoluble agaropectin acetate (30%) and the respective polysaccharides were isolated after deacetylation with alcoholic alkali. Agaropectin may also be separated by precipitation with cetylpyridinium chloride (Hjerten, 1962). Treatment of the supernatant with Fuller's earth binds the residual detergent and high yields of agarose may be obtained. However, the method is laborious as it involves frequent washings and centrifugation and consequently only small quantities can be fractionated (but see p. 154). Precipitation of agarose from hot solutions of agar can be achieved with polyethylene glycol 6000 (Carbo-wax 4000) (Russell *et al.*, 1964) and this allows fractionation of larger batches of material than the former method, but the yields are lower. Modification (Hegenauer and Nace, 1965) of the latter method has to some extent over-come this difficulty.

The proportion of agarose varies from species to species, agar from *Gelidium subcostatum*, *Pterocladia tenuis* and *Ceramium boydenii* comprising over 80% of agarose, whereas that from *Acanthopeltis japonica* contains only 28% (Araki, 1966). The average content seems to be of the order of 55 to 66%.

1. *Agarose*

Early structural studies were carried out on agar comprising both agarose and agaropectin although most of the results applied to the agarose fraction. Equal proportions of 2,4,6-tri-*O*-methyl-D-galactose and 2-*O*-methyl-3,6-anhydro-L-galactose were isolated from the hydrolysate of methylated agar

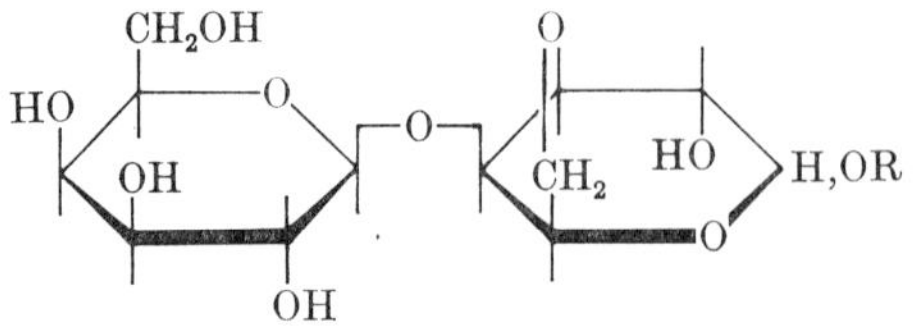

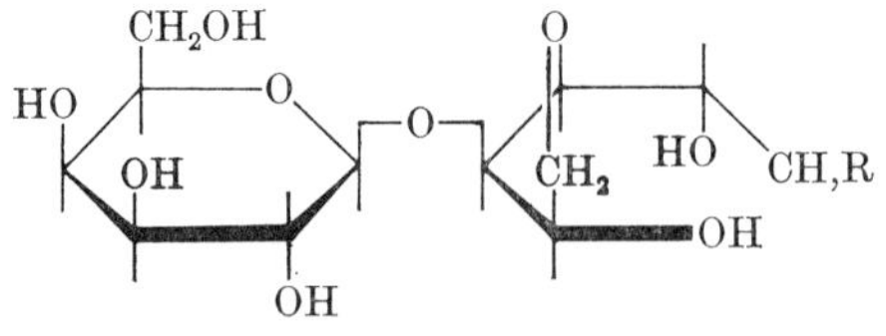

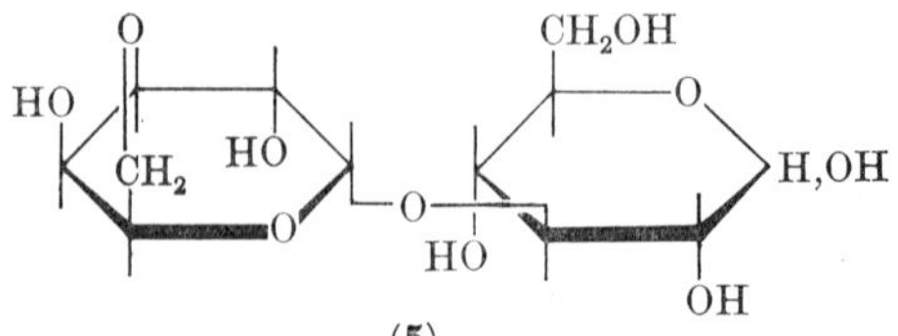

FIG. 1.

Plate LIX

Nature Printed by Henry Bradbury.

GELIDIUM corneum L. AMOUR.

Plate LVIII.

Nature Printed by Henry Bradbury.

GELIDIUM cartilagineum GAILL.

(Percival and Somerville, 1937; Forbes and Percival, 1939; Hands and Peat, 1938; Araki, 1937b, 1940, 1956, 1960) providing evidence that the polysaccharides are composed of 1,3-linked D-galactose and 1,4-linked 3,6-anhydro-L-galactose.

Agarobiose, the disaccharide, 4-O-β-D-galactopyranosyl-3,6-anhydro-L-galactose, $[\alpha]_D$ − 21°, has been isolated as the free sugar (1), as its diethyl dithioacetal (2) and as its dimethyl acetal (3) (Fig. 1), the last in 70% yield, by partial hydrolysis, mercaptolysis and methanolysis, respectively, of samples of commercial agar and of agar from *G. amansii* (Araki and Hirase, 1954) and from *Gracilaria confervoides* (verrucosa) (Clingman *et al.*, 1957). The latter authors calculated from their yield of methanolysed product that 76% by weight of the agar molecule is composed of agarobiose units. In more recent studies on the agarose fraction from *G. amansii*, an 82% yield of agarobiose as the dimethyl acetal (3) and glycoside (4) was separated after methanolysis (Araki and Hirase, 1960). The structure of this disaccharide has been confirmed by its synthesis from 1,2-O-isopropylidene 3,6-anhydro-L-galactose and acetobromo-D-galactose (Araki, 1966).

A second crystalline disaccharide, *neo*agarobiose, O-3,6-anhydro-α-L-galactopyranosyl (1 → 3)-D-galactose (5) (29%), $[\alpha]_D$ +20°, and a tetrasaccharide, O-3,6-anhydro-α-L-galactopyranosyl (1 → 3)-O-β-D-galactopyranosyl (1 → 4)-O-3,6-anhydro-α-L-galactopyranosyl(1 → 3)-D-galactose (4'-O-β-*neo*agarobiosyl-*neo*agarobiose) (40%) were isolated after partial enzymic hydrolysis by an enzyme extracted from the agar-digesting bacterium, *Pseudomonas kyotoensis* (Araki and Arai, 1956, 1957). By this preferential cleavage by acid of the α-1,3-links, and by the enzyme of the β-1,4-links, evidence is provided that the major structural feature of agarose is a linear chain of alternating 1,3-linked β-D-galactopyranose and 1,4-linked 3,6-anhydro-α-L-galactopyranose (Fig. 2, 6).

That 3,6-anhydro-L-galactose occurs at the non-reducing end and galactose at the reducing end of the molecule was deduced by Araki from the apparent absence of D-galactose or 3,6-anhydro-L-galactose in the products of enzymic hydrolysis and the absence of tetra-O-methylgalactose in the hydrolysate of methylated agar. A similar structure has been suggested for the unfractionated agar of *Gelidium cartilagineum* (O'Neill and Stewart, 1956), but with a half ester sulphate on about every tenth galactose unit.

Careful analyses by the Japanese School of agarose from different agarophytes (Hirase and Araki, 1961) have revealed the presence of 6-O-methyl-D-galactose together with small quantities of L-galactose and D-xylose (Araki, 1966) (see Table 1).

Although the quantity of 6-O-methyl-D-galactose varies from 0·8 to 20·8% in the different species, the sum of the D-galactoses is invariably 51 to 53% and the 3,6-anhydro-L-galactose and L-galactose about 46%.

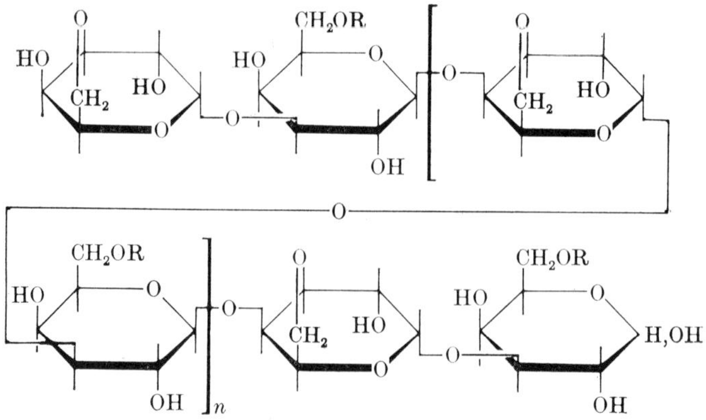

(6) R = H or CH$_3$

Agarose

Fig. 2.

Table 1   Percentage composition of different agaroses.

Source	D-Galactose	6-O-Methyl-D-galactose	3,6-Anhydro-L-galactose	L-Galactose	D-Xylose
Gelidium amansii	51·0	1·4	44·1	1·9	0·3
Gelidium subcostatum	45·1	7·3	43·8	1·8	1·3
Gelidium japonicum	50·8	1·6	44·0	1·9	0·3
*Pterocladia tenuis**	51·7	0·8	44·2	1·4	0·6
Acanthopeltis japonica	49·4	3·2	44·5	2·0	0·4
Campylaephora hypnaeoides	50·0	0·8	43·7	4·0	1·8
Gracilaria verrucosa	36·3	16·3	44·0	2·1	0·2
Ceramium boydenii	31·9	20·8	44·1	1·0	0·7

* See also Wu and Ho (1959).

A small quantity of 4-O-methyl-L-galactose has also been reported (Araki, 1967) as a constituent of *G. amansii* agar. It was isolated in crystalline form, but the quantity is so small it can have little structural significance apart from the fact that these particular units of L-galactose cannot be 1,4-linked.

That the 6-O-methyl residues are linked to C-4 of the 3,6-anhydro-L-galactose residues in the same way as the unsubstituted D-galactose units, was shown by the isolation of 6′-O-methylagarobiose dimethyl acetal from the methanolysate of the agar from *C. boydenii*. This disaccharide has been synthesized by condensation of 6-O-methyl acetobromo-D-galactose with 1,2-isopropylidene 3,6-anhydro-L-galactose (Araki, 1966).

Proposed General Formula for Agaroses [After Araki (1966)]

$$[\text{-3-D-Gal}(1 \rightarrow 4)\text{-L-Agal}(1]_x \rightarrow [3)\text{-D-Gal}(1 \rightarrow 4)\text{-L-Agal}(1]_y \rightarrow$$

$$\underset{\substack{| \\ 6 \\ \text{OMe}}}{}$$

Gal = galactose; Agal = 3, 6-anhydrogalactose; Me = methyl

Agarose from *G. japonicum*, for example, contains thirty-one times as much D-galactose as 6-*O*-methyl-D-galactose, hence for this species $x = 31$ and $y = 1$. It should be emphasized that this is a purely hypothetical formula since the relative positions of the 6-*O*-methyl-D-galactose and the D-galactose units are unknown and it is very improbable that they would be in respective halves of the molecule as depicted.

2. *Agaropectin*

Agaropectin, possibly a mixture of polysaccharides, comprises mainly D-galactose, 3,6-anhydro-L-galactose, some ester sulphate (3·5 to 9·7%) and D-glucuronic acid. Recent enzymic studies with a purified agarase from *P. atlantica* acting on agaropectin fractionated (Hjerten, 1962) from commercial Difco agar gives rise to a series of oligosaccharides (W. Yaphe, personal communication) which on electrophoresis in acetic acid-pyridine buffer (pH 7) yields three distinct bands. One of these is found on the starting line and comprises neutral sugars which, on thin layer chromatography, have mobilities identical with *neo*agaro-biose, -tetraose (major component) and -hexaose, a pattern roughly similar to that obtained from agarose. The two migrating bands appear from thin layer chromatography to consist of single sulphated fragments. Agaropectin clearly has many of the same structural features as agarose, and it is probable that with the advent of new fractionating techniques it will be shown to consist of a number of related polymers (cf., λ-carrageenan).

In addition, the agaropectins from *G. amansii* and *G. subcostatum* (Araki, 1966) contain about 1% of pyruvic acid. Hirase (1957) isolated a crystalline disaccharide consisting of substituted agarobiose after partial methanolysis of commercial agar and showed it to have the constitution: 4,6-*O*-1′ carboxy-ethylidene-β-D-galactopyranosyl-(1 → 4)-3,6-anhydro-L-galactose dimethyl acetal (**7**). It is this substituted galactose residue which yields pyruvic acid on acid hydrolysis.

An agaropectin-type of polysaccharide, $[\alpha]_D +40°$ has been isolated from *Ahnfeltia plicata* (Arai, 1961), but it differs from the agars discussed above in the presence of L-arabinose units. It comprises 3,6-anhydro-L-galactose, L-arabinose and D- + L-galactose in the molar ratios of 1:1·38:3·10, contains

(7)

D-glucuronic acid and 5·3% of ester sulphate, and gives a chloroform in-soluble acetate. In addition to the expected monosaccharides, a good yield of agarobiose [Fig. 1, (3)] and agarotetraose dimethyl acetals were isolated after partial methanolysis. After Haworth and Purdie methylations of this material, a fraction was separated (OMe, 33·3%; $[\alpha]_D$ − 86·3°) which was similar to the methylated agarose from *G. amansii* and which, on complete methanolysis, gave approximately equal molar proportions of the D- and L-sugars in the form of methyl 2,4,6-tri-*O*-methyl-D-galactopyranoside, and 2-*O*-methyl-3,6-anhydro-L-galactopyranoside and its dimethyl acetal. From these results the author concluded that the main repeating units of the polysaccharide are alternate residues of *β*-D-galactopyranose linked through C-1 and C-3 and 3,6-anhydro-*α*-L-galactopyranose connected through C-1 and C-4 as in agarose (see Fig. 2). No evidence was advanced for the part that the arabinose plays in the structure of this polysaccharide.

B. PORPHYRAN-TYPE POLYSACCHARIDES

Porphyran-type polysaccharides have been found in *Porphyra* and *Laurentia* spp. and in *Bangia fuscopurpurea*. Until the recent discovery of 6-*O*-methylgalactose as a constituent of agar, this methylated sugar, which is a common constituent of the porphyran-type of mucilage (Nunn and von Holdt, 1957; Su, 1958), was thought to distinguish the latter type of poly-saccharide from agar and from carrageenan. Now it appears that porphyran resembles agarose in containing 3,6-anhydro-L-galactose and 6-*O*-methyl-D-galactose and resembles carrageenan in containing galactose 6-sulphate (Turvey and Rees, 1961; Su and Hassid, 1962).

The galactan from *Porphyra umbilicalis* was shown to comprise 3,6-anhydro-L-galactose, L-galactose 6-sulphate, D-galactose, and 6-*O*-methyl-D-galactose (Peat *et al.*, 1961). It was found that these components vary widely from sample to sample at different seasons and from different environments (Rees and Conway, 1962a). Although it was found that 3,6-anhydro-L-galactose could vary from 5 to 19%, ester sulphate from 6 to 11%, 6-*O*-methyl-D-galactose 3 to 28%, and galactose from 24 to 45%; the sum of L-galactose 6-sulphate and 3,6-anhydro-L-galactose is always approximately equal to the sum of D-galactose and 6-O-methyl-D-galactose, and the galactose 6-sulphate accounts for nearly all the ester sulphate in the polysaccharide. At the same time, it has been shown that the galactose 6-sulphate in the polysaccharide from *P. umbilicalis* can be converted quantitatively into the 3,6-anhydride, chemically by treatment with alkali (Rees, 1961a) or enzymically by an extract from the seaweed (Rees, 1961b). It was found that of the 11·7% (as SO_4) ester sulphate in this sample, 86% was present as 6-sulphate.

Periodate oxidation and partial hydrolysis studies (Turvey and Williams, 1961, 1964) indicate that the D-galactose and its 6-O-methyl ether are linked through C-3 and L-galactose and the 3,6-anhydride through C-4. The following disaccharides were separated and characterized from a partial hydrolysate:

β-D-Gal(1 → 4)L-Gal6S;   6-O-Me-β-D-Gal(1 → 4)L-Gal6S;

α-L-Gal6S(1 → 3)D-Gal;   L-Gal6S(1 → 3)6-O-Me-D-Gal

Gal = galactose ; S = ester sulphate ; Me = methyl

These linkages were confirmed and the essential similarity of porphyran to agarose emphasized by its conversion into methylated agarose (Anderson and Rees, 1965). The L-galactose 6-sulphate units were converted into the 3,6-anhydro-derivative by alkaline elimination and the product was methylated. The methylated "alkali-modified" porphyran had sulphate (as SO_3'') 1·5% and $[\alpha]_D$ − 79° (cf., methylated agarose $[\alpha]_D$ − 93°). Furthermore, on hydrolysis it gave 2,4,6-tri-O-methyl-D-galactose and 3,6-anhydro-2-O-methyl-L-galactose as the major products, and partial methanolysis gave a 62% yield of derivatives of methylated agarobiose, 3,6-anhydro-2-methyl-4-O-(2,4,6-tri-o-methyl-β-D-galactopyranosyl)-L-galactose (8).

(8)

Correcting for the amount of cleavage of the glycoside when subjected to methanolysis, the total yield of disaccharide in the polysaccharide rose to 82%, and when allowance was made for the ester sulphate still present in the methylated "alkali-modified" porphyran, this became 90% for the repeating unit content of the non-sulphated part of the molecule. The high yield of this disaccharide proves that the major part of the macromolecule consists of alternating units of 1,3-linked D-galactose and 1,4-linked 3,6-anhydro-L-galactose or L-galactose 6-sulphate. It must be remembered, however, that these conclusions are based on a 75% yield of "alkali-modified" methylated material. Assuming that the molecules that survived methylation are representative of the native material, it may be concluded that porphyran consists of a linear chain of alternately 3-linked β-D-galactosyl and 4-linked α-L-galactosyl units, the α-linkage being inferred from the optical rotation of the methylated material (see Fig. 3). In this case the simple sequence is hidden by the fact that some of the D-galactose is methylated at C-6 and that the L-galactose is present both as 3,6-anhydride and 6-sulphate. No conclusive evidence for the site of the small proportion of additional sulphate has yet been presented.

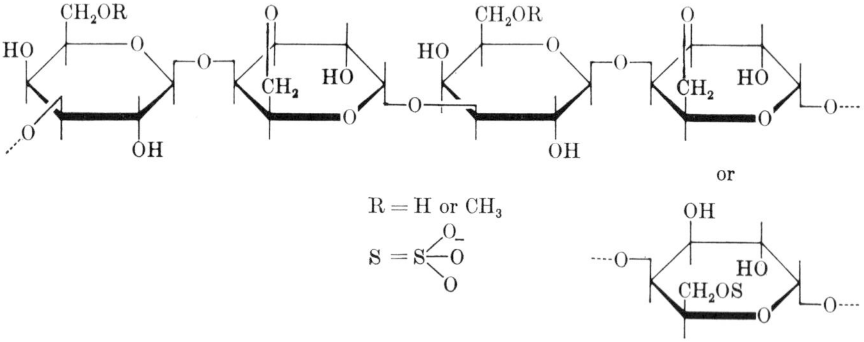

FIG. 3.

Enzymic studies have confirmed some of these findings. An agarase from a marine bacteria, a *Cytophaga* sp. grown on a culture medium containing porphyran as the sole carbon source, cleaved the porphyran into D-galactose, 6-*O*-methyl-D-galactose, the disaccharide, *neo*agarobiose [Fig. 1, **(5)**] and higher homologues, and a tetrasaccharide to which the constitution:

3,6-Agal(1 → 3)6-*O*-Methyl-D-Gal(1 → 4)-3,6-Agal(1 → 3)-D-Gal

Agal = 3,6-anhydro-L-galactose

has been tentatively assigned. In addition, a number of sulphated oligosaccharides, but no monosaccharide sulphates, were present in the hydrolysate. The enzyme appears to dislike 6-*O*-methylgalactosyl linkages since only traces of the disaccharide, 3,6-anhydrogalactosyl (1 → 3)-6-*O*-methyl-D-galactose could be detected (Christison and Turvey, 1966).

Similar galactans are synthesized by a variety of *Porphyra* species (Nunn and von Holdt, 1957; Rees and Conway, 1962a), by *Bangia fuscopurpurea* (Wu and Ho, 1959) and by *Laurencia pinnatifida* (Bowker, 1966). All attempts to fractionate the mucilage from the latter weed were unsuccessful although evidence of its polydispersity was obtained. It contains mainly D- and L-galactose, 6-O-methyl-D-galactose, 3,6-anhydro-L-galactose, D-xylose and ester sulphate in the approximate molar ratios of 2·5:0·85:1·65:3·5:1·0:3·5 together with about 6% of galacturonic acid. Hydrolysis, mercaptolysis and methylation studies established also the presence of 2-O-methyl-L-galactose and L-galactose 6-sulphate and led to the tentative identification of 3,6-anhydro-2-O-methyl-L-galactose and of D-galactose 2-sulphate. Evidence for the presence of 6-O-methyl-D-galactose monosulphates and for additional branching points in the molecule was obtained, but the relationship with the xylose and galacturonic acid residues was not elucidated. From the available evidence the author suggests that all the D-galactose units are 1,3-linked and all the L-galactose residues 1,4-linked. Although the alternating sequence was not established, the D:L molar ratio of 1·0:1·2 for all the galactose units is suggestive of the basic repeating unit characteristic of the galactans already discussed, although this is largely masked in the present mucilage by ester sulphate and methoxyl groups. Furthermore, the presence of 2-O-methyl-galactose derivatives distinguish the mucilage from porphyran and agar. Treatment of the mucilage with alkali indicates that L-galactose 6-sulphate and 2-O-methyl-L-galactose 6-sulphate are converted into the respective 3,6-anhydrogalactoses and the author considers that the former are the biological precursors of the anhydro sugars.

C. Carrageenan-type Polysaccharides

The name carrageenan was originally applied to the polysaccharide from species of *Chondrus* and *Gigartina* but compounds of similar structure are found in plants of the genera *Furcellaria, Eucheuma, Hypnea, Iridaea* (= *Iridophycus*) and *Polyides*. The polysaccharides are extracted from the algae with hot water, but unlike agar they cannot be purified by freezing; they are normally isolated by precipitation from the aqueous extract by the addition of alcohol.

The carrageenan-type polysaccharides differ chemically from agar in that 3,6-anhydro-α-D-galactose takes the place of the anhydro-L-sugar in agar (Percival, 1954; O'Neill, 1955a), and they have a higher content (about 24%) of mainly alkali stable ester sulphate (Buchanan *et al.*, 1943; Dewar and Percival, 1947; Johnston and Percival, 1950).

Like agar, carrageenan can be fractionated into two polysaccharides κ- and λ-carrageenan. In this case, separation is achieved by precipitation of the κ-carrageenan from a 0·25% aqueous solution with dilute potassium chloride solution to a 0·25M concentration (Smith and Cook, 1953). Again like agar, the proportion of κ- and λ-carrageenan varies in the different seaweeds and, in the case of carrageenan, with the season and habitat of a single species (Black *et al.*, 1965). The ratio of κ to λ in *Chondrus crispus* carrageenan varies from 0·7 for seaweed growing during July/August on Ilo Grande, Cotes du Nord, France to 2·1 for the same species growing in the Northumberland Strait, Pictou Co., Nova Scotia, and it rises to 3·05 for the latter site in November. At the same time carrageenan from this species does not always fractionate sharply. Whereas *Gigartina stellata* and *G. radula* had similar κ-carrageenan contents to that of *Chondrus*, *G. acicularis* and *G. pistillata*, both from Portugal, had a low κ content, but it must be emphasized that this might be due to seasonal and/or habitat factors. Fractional precipitation of extracts from *C. crispus*, *G. stellata* and *G. skottsbergii* with increasing concentrations of potassium chloride (Pernas *et al.*, 1967) indicate that more than two components can be separated by this means.

Neither *Eucheuma spinosum* nor *Polyides rotundus* carrageenans can be fractionated with potassium ions. The former accounts for 55% of the dry weight of the plant and, apart from a higher ester sulphate content, structurally resembles κ-carrageenan. The latter appears to be devoid of 3,6-anhydrogalactose residues.

An enzyme, κ-carrageenase, from *Pseudomonas carrageenovora* (which is a specific hydrolase for κ-carrageenan) has also been used to determine the κ content of about fourteen different species of red seaweeds (Yaphe, 1959). The results are not in complete agreement with those from the chemical fractionation, for example, *G. radula* was found to be low in κ-carrageenan. However, this only serves to emphasize the need for seasonal and habitat factors to be taken into account before comparisons between different species can be made. It is interesting that carrageenase revealed a higher content for the κ fraction in *Hypnea musciformis* and in a *Yatabella* sp. than in *C. crispus*.

1. *Structural Investigations of Chondrus and Gigartina Carrageenan*

2,6-Di-O-methyl-β-D-galactose was isolated as the major sugar from the hydrolysate of methylated carrageenan from *G. stellata* (Dewar and Percival, 1947). In view of the alkali stability of the ester sulphate groups (see p. 46), it was concluded therefore that 1,3-linked galactose units sulphated at C-4 constitute a major structural feature of the polysaccharide.

κ-*Carrageenan.* κ-Carrageenan, the fraction of carrageenan precipitated with potassium chloride, was shown (O'Neill, 1955b) to consist of D-galactose (about six parts), 3,6-anhydro-D-galactose (about five parts) and ester sulphate groups (seven parts). O'Neill separated in good yield the disaccharide, carrabiose as the dithioacetal, 4-O-β-D-galactopyranosyl-3,6-anhydro-D-galactose diethyl dithioacetal (**9**) from the mercaptolysate of this fraction of carrageenan.

It was found that treatment of κ-carrageenan with warm alkali (Rees, 1961a) removes a small proportion of the ester sulphate with concomitant

Plate XI

CHONDRUS Crispus, LYNGB.

Nature Printed by Henry Bradbury

Facing page 138

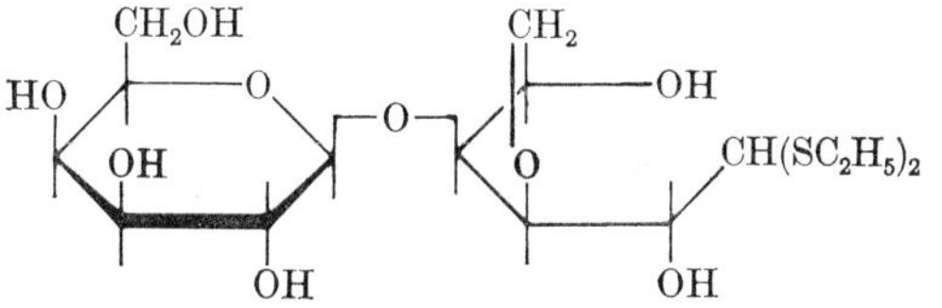

(9)

formation of 3,6-anhydrogalactose, indicating the presence of some 1,4-linked galactose 6-sulphate.

More recent enzymic studies (Weigl et al., 1966) with a κ-carrageenase from *Pseudomonas carrageenovora* (Weigl and Yaphe, 1966) have led to the separation of an homologous series of oligosaccharides (80%), based on the disaccharide, *neo*carrabiose sulphate (10), and 20% of an enzyme resistant fragment. The disaccharide, $[\alpha]_D$ +179°, was characterized as O-α-3,6-anhydro-D-galactopyranosyl(1 → 3)D-galactopyranose 4-sulphate (10), indicating that the enzymic hydrolysis occurs at the β-1,4-linkage (cf., enzymic hydrolysis of agar p. 131).

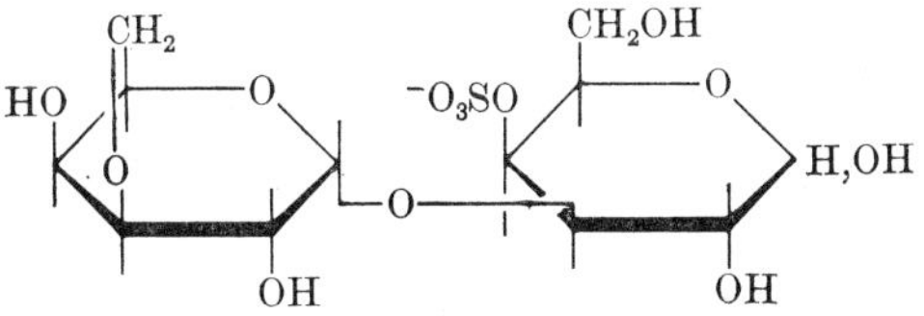

(10)

The resistant fragment contains a higher proportion of sulphate than κ-carrageenan and comprises a molar ratio of galactose:3,6-anhydrogalactose:sulphate of 1·4:1:1·8. Furthermore, on treatment with alkali, 19% of the ester sulphate was removed with formation of an additional 14% of 3,6-anhydrogalactose and a change of the infrared spectra to that of κ-carrageenan. The alkali-modified material was found to be susceptible to further enzymic hydrolysis by κ-carrageenase. The authors conclude, therefore, that the enzyme resistant region is a biogenetically "unfinished" portion of the κ-carrageenan molecule which still contains galactose 6-sulphate units.

Methylation studies (Dolan, 1965) have confirmed that practically all the 3-linked units occur as D-galactose sulphated at C-4 and have shown that the 4-linked 3,6-anhydro-D-galactose units are partially sulphated at C-2. A small proportion of 1,4-linked D-galactose units was found to be either sulphated at C-6 or disulphated at C-2 and C-6, and it is these units which give rise to the anhydrosugar under the action of alkali. These results are supported by paper chromatographic identification of galactose 2-, 4- and 6-sulphate in partial acid hydrolysates of κ-carrageenan (Painter, 1966).

In order to establish that the 1,3- and 1,4-linkages are in alternation in this polysaccharide, mild methanolysis before and after alkali-modification was carried out (Anderson and Rees, 1966), and the yields of carrabiose dimethyl acetal measured. Parallel experiments on carrabiose dimethyl acetal gave a measure of the loss due to side reactions. When allowance was made for these, an 88% and 99% *"carrabiose content"* which is defined here as *the proportion of 3,6-anhydrogalactose combined in the form of carrabiose units which is readily released by mild methanolysis*, was obtained for κ-carrageenan and the "alkali-modified" material, respectively. Bearing in mind this definition, the increase in the "carrabiose content" on alkali treatment must result from 6-sulphated galactose residues present in place of 3,6-anhydrogalactose in the alternating sequence in the κ-carrageenan and cannot be the result of contamination with λ-carrageenan (see p. 143) since a mixture of a perfect carrabiose polymer with λ-carrageenan could not account for this change in "carrabiose content". The alternating sequence of galactose units in κ-carrageenan are shown in Fig. 4 together with the residues (**11**) which occasionally change the form of the 1,4-linked units.

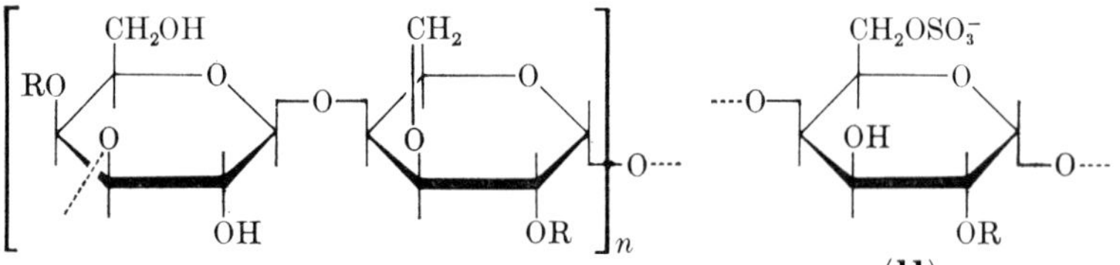

R = H or SO_3^-

FIG. 4.

λ-Carrageenan. λ-Carrageenan, the material remaining in the supernatant after potassium chloride fractionation of carrageenan, is a mixture of highly sulphated galactans (sulphate about 35%), comprising D-galactose, a small proportion of 3,6-anhydro-D-galactose and a little (about 2%) of L-galactose. Fractional precipitation of λ-carrageenan with alcohol yields polymers as major fractions consisting solely of D-galactose and minor fractions containing in addition L-galactose, glucose and xylose (Smith *et al.*, 1954). The last two sugars are considered to arise from contaminating floridean starch and a xylan. Samples of American λ-carrageenan from *C. crispus* were found to be practically devoid of 3,6-anhydrogalactose, whereas three samples of British λ-carrageenan from *G. stellata* contained appreciable amounts.

Partial acetolysis of λ-carrageenan led to the isolation of a high yield of 3-O-α-D-galactopyranosyl-D-galactose (**12**) and the "supposed" corresponding trisaccharide (Morgan and O'Neill, 1959) evidence that a major structural feature is α-1,3-linked galactose units, but indication of other structural

(12)

features was revealed by the reduction of periodate by this polymer which showed that one in every ten units contained a glycol grouping.

Evidence that galactose 6-sulphate residues with a free hydroxyl group on C-3 were present in the λ-fraction, as well as in the κ-fraction, was deduced from the increase in 3,6-anhydrogalactose units from 1·6% to 14% (Rees, 1963) (and the reduction of sulphate by about one third) on treatment of a commercial sample of λ-carrageenan (sulphate about 27%) with warm alkali (cf., Porphyran p. 135). The product is hereinafter known as "alkali-modified" λ-carrageenan. That the sulphate was indeed linked to C-6 was confirmed by the characterization of galactose 6-sulphate as one of the products of partial hydrolysis of λ-carrageenan.

Comparative methylation analysis (Dolan and Rees, 1965) of unmodified λ-carrageenan, and of products obtained by desulphation and by alkali treatment, revealed that the 1,3-linked units are mainly sulphated at C-2 with fewer sulphated at C-4 and that some are sulphate free.

Partial mercaptolysis of the "alkali modified" λ-carrageenan gave about 6·5% yield of carrabiose diethyl dithioacetal (9) (Rees, 1963); evidence that some of the galactose 6-sulphate units in the native polysaccharide are β-1,4-linked.

In view of the low reduction of periodate by λ-carrageenan and its high sulphate content, it seemed probable that these latter units, which would be cleaved by periodate, are rendered immune to this reagent by further sulphation. Proof of this was obtained by the isolation of 3,6-anhydro-galactitol 2-sulphate (13) from the "alkali modified" λ-carrageenan after mild acid cleavage of the 3,6-anhydrogalactosyl linkages followed by reduction and Smith degradation (Fig. 5). In the native polysaccharide these units are present as 1,4-linked D-galactose 2,6-disulphate.

The infrared absorption spectrum of λ-carrageenan (Rees, 1963) (a very broad band at 810 to 860 cm^{-1} with a maximum at 827 cm^{-1}) supports the presence of all three types of ester sulphate, the 2-sulphate being equatorially disposed in the C1 conformation, the 4-sulphate being axially disposed, and the 6-sulphate being attached to a primary alcoholic group. This is further supported by the reasonably sharp band at about 850 cm^{-1} revealed by "alkali-modified" λ-carrageenan consistent with loss of the 820 band due to C-6 sulphate leaving a molecule with a higher proportion of 4-sulphate.

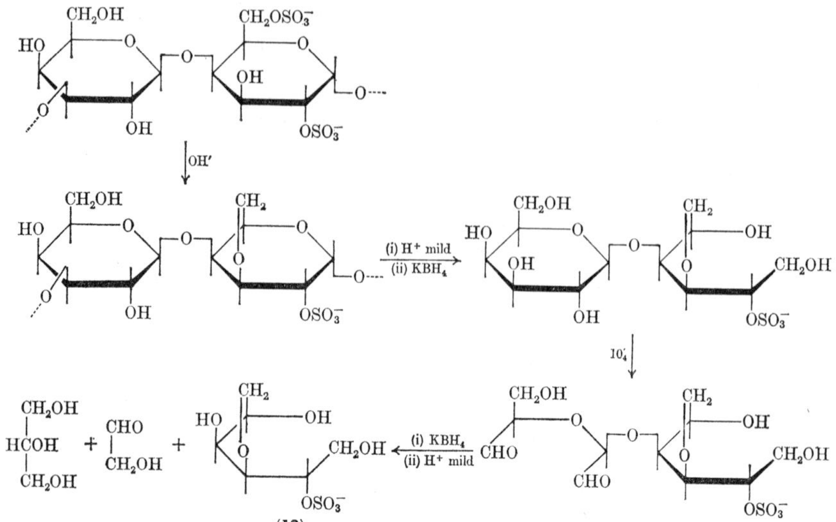

FIG. 5.

"Alkali-modified" λ-carrageenan as its potassium salt can be separated from a "third" component (Dolan, 1965; Anderson and Rees, 1966). The latter, depending on the sample, may be a minor or major constituent of the mixture. It was shown by methylation studies of "alkali-modified" λ- and "third" component that all the 3-linked galactose 2-sulphate occurs in the former, and all the 4-sulphate in the latter. Dolan and Rees (1965) propose, therefore, that the name λ-carrageenan should be reserved for the material which is devoid of 4-sulphate and of 3,6-anhydrogalactose (Fig. 6). In addition it should be emphasized that the proportion of 1,4-linked units sulphated at C-2 is greater in the λ-carrageenan than in the "third" component.

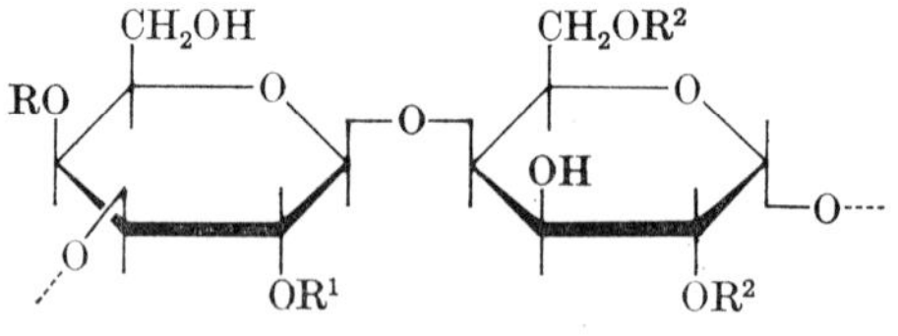

FIG. 6. λ-Carrageenan R = H; R' = SO_3^- or H; R^2 = SO_3^- or H
Third component R = SO_3^- or H; R' = H; R^2 = SO_3^- or H

In view of the uncertainties of fractional precipitation (p. 138) the present writers suggest also that the name κ-carrageenan should be reserved for the polymer with the structure described on p. 140.

Mild methanolysis of "alkali modified" λ-carrageenan and "alkali-modified" "third" component, as for the κ-fraction, gave a "carrabiose content", calculated as the proportion of the 3,6-anhydro units which appear as carrabiose, of 94·5 and 96%, respectively. It appears at first sight that these fractions are not therefore based on the perfect alternation of $1 \to 3$ and $1 \to 4$ linkages. However, if we consider, for example, a possible fragment of either of these molecules to be:

$\to 3Gal(1 \to 4)Agal(1 \to \vdots \to 3)Gal(1 \to 4)Gal(1 \to 3)Gal(1 \to 4)Agal(1 \to \vdots \to 3)Gal(1 \to 4)Agal(1 \to$

then mild methanolysis would only cleave it at the points shown with one molecule of 3,6-anhydrogalactose held in a tetrasaccharide. The 1,4-linked galactose may be present in the native polymer as such or it may have arisen from the 6-sulphate during alkali treatment. From this it is clear that it is possible to have a perfect alternation of 1,3- and 1,4-linkages without a 100% yield of "carrabiose content" on mild methanolysis. This formulation does not permit the presence of the "supposed" 1,3-linked trisaccharide, separated after acetolysis, as a structural feature, but its identification was only tentative and further work is in progress to completely characterize this oligosaccharide.

It seems, therefore, that all the fractions of carrageenan are based, like agar and porphyran, on an alternating sequence of 1,3- and 1,4-linked galactose units, the fractions differing from one another only in their proportions of 3,6-anhydro-sugar and sulphate, and in the various sites of the sulphate groups.

The galactan of *Chondrus Ocellatus* Holmes, $[\alpha]_D +54°$, is also of the carrageenan-type (Araki and Hirase, 1956). It contains 24% of sulphate and 1,3-linked β-galactose units. In addition, partial methanolysis gave a 73% yield of carrabiose derivatives proving that the major part of the polysaccharide comprised alternate units of 1,3-linked β-D-galactose and 1,4-linked 3,6-anhydro-α-D-galactose, the α-linkage of the anhydrogalactose being inferred from the positive rotation of the polysaccharide.

2. Structure of Other Carrageenan-type Polysaccharides

Furcellaria fastigiata, a seaweed particularly abundant in the Kattegatt, is the source of Danish agar, furcellaran. It has $[\alpha]_D +75°$ and comprises 43 to 46% D-galactose, 30% 3,6-anhydro-D-galactose and about 20% ester sulphate and a trace of xylose (Clancy *et al.*, 1960; Painter, 1960). Partial mercaptolysis gave diethyl mercaptals of D-galactose, 3,6-anhydro-D-galactose and carrabiose (9) (42%), and enzymic hydrolysis indicated the presence of 56% of κ-carrageenan-like material (Yaphe, 1959). Although the polysaccharide has many similarities with carrageenan, including the fact that it gels in the presence of potassium chloride, it differs in that only two in five residues are sulphated; of these a small proportion is converted into 3,6-anhydrogalactose on warming with alkali (Rees, 1961a). Chromatographic analysis of partial acid hydrolysates of laboratory and commercial samples of furcellaran (Painter, 1966) has led to the identification of C-2, C-4 and C-6 galactose monosulphates confirming the essential similarity of this material with κ-carrageenan.

After simultaneous deacetylation and methylation of this extract, a 15% yield of methylated material with a negative rotation was recovered. This gave only 2,3,4,6-tetra- 2,4,6-tri- and 2,4-di-O-methylgalactoses on hydrolysis, and was considered to represent a more resistant portion of the molecule consisting mainly of 1,3-linked β-D-galactose units. It seems that the polysaccharide is less highly organized on an alternating basis than κ-carrageenan, but has occasional small groups of D-galactose in the chain. On the other hand, the excess galactose may originate from a contaminating galactan which escaped detection in the ultracentrifuge (Painter, 1960) and a decision on this must await further study.

Chemical examination of the sulphated polysaccharides of *Hypnea specifera* (Clingman and Nunn, 1959) and *Eucheuma* sp. (O'Colla, 1962) indicated their essential similarity to κ-carrageenan, and enzymic hydrolysis (Yaphe, 1959) supported the presence of 87% of κ-carrageenan-like material in *H. musciformis*. However, the infrared spectra of *H. specifera* and *Eucheuma* polysaccharides differed in some respects from that of κ-carrageenan, but as with κ-carrageenan, treatment with warm alkali was found to remove a small proportion of ester sulphate with the simultaneous formation of 3,6-anhydrogalactose (Rees, 1961a).

The presence of ester sulphate at C-6 of the galactose units had early been reported for the polysaccharide from *Iridaea laminarioides* (= *Iridophycus flacidium*) (Hassid, 1933, 1935; Mori, 1950) together with the presence of α-1,3-linked galactose units. By enzymic hydrolysis Yaphe (1959) found 36% of κ-carrageenan in this material.

D. OTHER SULPHATED GALACTANS

A sulphated polysaccharide (funoran; Japanese Fukuofunori), $[\alpha]_D$ $-20 \cdot 6°$, which showed similarities with both agar and carrageenan, is synthesized by *Gloiopeltis furcata*. It resembles carrageenan in its high sulphate content (18·5%) and the fact that it gives a highly viscous solution but does not gel. It is composed mainly of D-galactose and 3,6-anhydro-L-galactose and after partial methanolysis, 50% of the methanolysate was separated as crystalline agarabiose dimethyl acetal [Fig. 1 (**3**)] indicating a similarity with agar (Hirase *et al.*, 1956, 1958). In addition to the agarobiose, methyl D-galactoside,

3,6-anhydro-L-galactose dimethyl acetal and about 2% of methyl xyloside were also separated. The latter sugar is considered to be possibly derived from a contaminating xylan. This polysaccharide has not yet been examined for methylated galactoses.

The galactan sulphate from the calcareous red alga, *Corallina officinalis* contains both D- and L-galactose in the molar ratio of 1·3:1 and one sulphate group for every three to four galactose units (Turvey and Simpson, 1966). It appears to be built up on the same general pattern since it contains equal quantities of 1,3- and 1,4-linked galactose units. It is devoid of 3,6-anhydro-galactose, but contains ester sulphate linked to C-6 of the L-galactose units and to C-4 of either D- or L-galactose.

A highly branched sulphated galactan which, by Barry degradation (see p. 41), was shown to contain a 1,3-linked galactose back-bone, has been extracted from *Dilsea edulis* (Barry and McCormick, 1957). In addition, xylose and glucuronic acid and a trace of 3,6-anhydrogalactose are present, and some of the galactose units are sulphated at C-4. The glucuronic acid and 3,6-anhydride are considered to be present as end groups and 1,4-linked galactose and 1,3-linked xylose occur as side chains.

More recent work (Rees, 1961a) has indicated that in addition to galactose 4-sulphate, small amounts of 1,4-linked 6-sulphate are present and these are converted into 3,6-anhydrogalactose on treatment of the polysaccharide with alkali. Furthermore, the xylose units are considered to arise from a contaminating xylan. A more highly sulphated galactan of this type is synthesized by *Dumontia incrassata* (Dillon and McKenna, 1950a, 1950b).

A very complex mucilage has been extracted from *Polysiphonia fastigiata* (Peat and Turvey, 1965) which contains both D- and L-galactose, 6-O-methyl-D- and L-galactose 3,6-anhydro-D- and L-galactose and possibly D- and L-galactose 6-sulphate. Methylation provided evidence of 1,3-linked galactose units.

D-Xylose and D-glucuronic acid have been reported in a number of these mucilages but in no case has unequivocal evidence been advanced that they are not derived from a contaminating xylan and glucuronan. A final decision on this awaits further study.

Although for convenience these galactans, which are unlike any other natural polymers so far investigated, have been divided into three types, it is gradually emerging as knowledge of their structure increases that it is better to regard them as a single family of polysaccharides. Some differences between species within these arbitrary classes are almost as great as those between typical examples of the classes. For example, the galactans fractionated from *Gracilaria verrucosa* and *Ceramium boydenii* classed by Araki (1966) as agaroses (see Table 1) are, in respect to their 6-O-methyl-D-galactose contents of 16·3% and 20·8%, closer to porphyran than to the agar from *Gelidium amansii*, while on the basis of their very low sulphate contents they certainly belong among the agaroses. The structural units characterized in the different galactans are given in Table 2.

The galactans that have been most extensively investigated have been shown to consist of chains of alternate 1,3- and 1,4-linked D- and L-galactose units. The 1,4-linked units are often present as the 3,6-anhydride and the simple alternation is frequently masked by the presence of ester sulphate and methoxyl groups. Further study of the mucilages, which on preliminary investigation appear to deviate from this strictly alternating sequence, may reveal the presence of small proportions of contaminating polysaccharides which mask this alternating pattern.

III. MOLECULAR WEIGHTS

The properties of the galactans of the Rhodophyceae, for example the high viscosity of their solutions and the formation of gels, indicates that they are highly polymerized, but determinations of molecular weight have been made only for some of the compounds of commercial importance. Measurements of the osmotic pressure of fractions extracted by water from electrodialysed agar showed that they had molecular weights in the range 5000 to 30,000 (Morozov, 1935; Lipatov and Morozov, 1935). Figures of 160,000 and 110,000 have been obtained for laboratory extracted and commercial agar (Selby and Selby, 1959).

The κ- and λ-fractions of commercial carrageenans, as well as *Chondrus* extracts, made at different temperatures have been studied by ultracentrifuge and viscosimetric methods (Smith *et al.*, 1954, 1955; Goring and Young, 1955). A comparison of the κ- and λ-fractions from the same sample of carrageenan showed that the λ-fraction had a higher molecular weight. Figures obtained for κ-fractions ranged from 260,000 to 320,000 and for λ-fractions from 330,000 to 790,000. In all cases further fractionation into products with different molecular weights could be brought about by precipitation with increasing concentrations of alcohol (Smith *et al.*, 1955). Samples extracted at different temperatures had molecular weights ranging from 300,000 to 1,400,000. Higher temperatures were necessary to extract the larger molecular species, but prolonged heating led to degradation and the most highly polymerized material was obtained by a short extraction at 100°.

IV. ENZYMIC HYDROLYSIS

An enzyme κ-carrageenase, which will hydrolyse the β-1,4-linkages in κ-carrageenan (see p. 139) has been precipitated from the cell-free culture medium of *Pseudomonas carrageenovora*, with ammonium sulphate (Weigl

Table 2   Residues found in Rhodophyceae Galactans.

Extractive	Galactose and derivatives							Sulphate on			Other units
	Galactose		3,6-Anhydro		6-O-Methyl	2-O-Methyl	4-O-Methyl				
	D	L	D	L	D	L	L	C-2	C-4	C-6	
Agarose from a variety of genera (see Table 1)	**[1]	*		**[1]	*[1]		*				Xylose (trace) Pyruvate[3]
Agaropectin from several genera	**			**	**				*[2]		Glucuronic acid
Porphyran from several Porphyra sp.	***[4]	*[4]		*[4]	*[4]					*[4]	
From Laurencia pinnatifida	**	*		**	*	*		*		*	Xylose Galacturonic acid
Carrageenan from Chondrus and Gigartina spp.											
κ-fraction	**	*	**					*	*	*	
λ-fraction	**		*					*	**	**	
Third Component	**		**						**	**	
Furcellaran	**							*	*[2]	*	
Fumoran	**			**				*[2]			

Notes: ** Constitutes more than 20% of the galactan.

* Present, but generally less than 20%.

[1] Sum of D-galactose and 6-O-methyl-D-galactose approx. equal to 3,6-anhydro-L-galactose.

[2] Position of sulphate groups not established.

[3] In Gelidium amansii and G. subcostatum only.

[4] Sum of D-galactose and 6-O-methyl-D-galactose and D-galactose 6-sulphate approx. equal to sum of 3,6-anhydro-L-galactose and L-galactose 6-sulphate.

and Yaphe, 1966). The precipitate also contains a λ-carrageenase which can be inactivated by incubation for 2 hr at 35° (Weigl *et al.*, 1966). The κ-carrageenase was purified by dissolution and dialysis, followed by chromatography on hydroxyapatite, then heated to inactivate the λ-carrageenase. This was followed by chromatography on DEAE-cellulose and electrophoresis on Gelman cellulose acetate at pH 7·5, and on starch gel in *tris*borate buffer at pH 8·7. The enzyme is stable at room temperature and can be stored at − 20°. It does not degrade λ-carrageenan or agar.

A crude agar degrading enzyme, agarase, was separated by Araki and Arai (1956) from *P. kyoteonsis* and utilized, in their structural studies, to cleave the β-1,4-links in agarose (see p. 131). Yaphe (1957) has also isolated a polysaccharase from *P. atlantica* and has shown this to be capable of hydrolysing agar-type polysaccharides, both agarose and agaropectin, but not κ-carrageenan. It appears to be specific for the 3,6-anhydro-α-L-galactopyranosyl(1 → 3)-D-galactose repeating unit (Yaphe, 1966).

When *P. atlantica* was grown on synthetic seawater with purified commercial agar, a maximum production of agarase occurred in between 20 to 24 hr, and the enzyme was precipitated from the cell-free culture medium on addition of ammonium sulphate to 70% saturation. Dialysis of the dissolved precipitate to an ionic strength of less than 0·1 was followed by adsorption of the enzyme on to carboxymethylcellulose. The enzyme was eluted with a gradient phosphate buffer pH 6·9 form 0·01 to 0·1 and rechromatographed on carboxymethylcellulose, dialysed and freeze-dried. The purified enzyme has a sharp peak at 280 mμ, gives a single band on electrophoresis and has a broad pH optimum between 5·0 and 7·5. Dilute solutions are rapidly inactivated but can be stabilized for at least 24 hr at 4° by addition of 0·1 mg/ml of Bovine serum albumin. Although heat labile, the enzyme can be protected from heat denaturation by the presence of its substrate. It hydrolyses agar with a rapid decrease in viscosity and a slow increase in reducing sugar, indicating an attack on internal linkages. The same series of homologous oligosaccharides as those obtained by the Japanese workers but starting from *neo*agarotetraose was found to be present. Incubation with fresh enzyme degrades the higher oligosaccharides to give mainly *neo*agarotetraose. This can be split into galactose, 3,6-anhydrogalactose and *neo*agarobiose on incubation with an intracellular enzyme preparation from *P. atlantica*. Apparently two intracellular enzymes are present, one of which cleaves β-1,4-linkages and the other α-1,3-linkages of the tetrasaccharide.

An enzymic extract from *Cytophaga* sp., a bacterium isolated from seawater, partially degraded porphyran and agarose, but had little or no action on carrageenan and structurally similar galactans (Christison and Turvey, 1966). Prolonged action of the enzyme preparation on porphyran gave 30% of aqueous ethanol-soluble carbohydrates comprising D-galactose,

6-O-methyl-D-galactose, *neo*agaro-biose, -tetraose, and higher saccharides similar to the products from agarose. About 70% of the porphyran was recovered as high molecular weight material.

A carbohydrase fraction from the viscera of *Turbo cornitus* and some other gastropods (Hirase *et al.*, 1956, 1958) hydrolyses *Chondrus ocellatus* mucilage without producing any inorganic sulphate. It is unstable in the dry state but may be stored for a limited period at 0°. The optimum pH and temperature were found to be 4·0 and 40°.

V. PROPERTIES

The galactans either dissolve in cold water to give viscous solutions, or dissolve only on heating, in which case gels are formed on cooling. As might be expected from the wide range of composition of the galactans from different genera of algae, the viscosities of the solutions and the melting and setting temperatures of the gels vary considerably as do the physical characteristics of the gels, such as elasticity and rupture strength.

With many of the sulphated galactans, gel formation is dependent on the nature of the cations combined with the half ester sulphate groups and on other salts present in the solution.

A gel can be regarded as a network of precipitated polymeric material, holding within it water and material which has remained in solution, so that the gel strength is determined to a large extent by the amount of polymer which has come out of solution to form the network. Changes of temperature which affect the solubility of the polymer, and the presence or absence of substances which will cause precipitation without excessive dehydration, can therefore determine whether a gel is formed and how strong it will be.

The effect of different constituent groups on the solubility of the polymer in water and its sensitivity to different precipitating agents, is therefore likely to give a useful basis for studying the relation between the physical properties of the different galactans and their chemical composition. Painter (1966) has pointed out that 3,6-anhydrogalactose is a much less hydrophilic compound than galactose, and its influence on the solubility of the polysaccharides is illustrated by the fact that agarose, which has a high content of 3,6-anhydrogalactose, is insoluble in cold water and yields very strong gels (Hjerten, 1964). On the other hand, the presence of sulphate groups makes the galactans more soluble; κ-carrageenan with more sulphate than furcelleran is more soluble in cold water, but the 3,6-anhydro content of these two polysaccharides renders them less soluble than, for example, λ-carrageenan.

From a consideration of chromatographic mobilities of salts of D-galactose sulphates and the hydration numbers of the cations, it was concluded that the potassium salts are less hydrophilic than sodium and barium salts of these

6+

compounds (Painter, 1967). The addition of a potassium salt to a dissolved sulphated galactan which is near to its solubility limit at that temperature could, therefore, bring about its precipitation. This is in line with the effect of potassium salts on κ-carrageenan and similar galactans in raising the setting and melting temperature of their gels, increasing gel strength, and allowing the fractionation from λ-carrageenan. A combination of the effect of both the potassium salt and of 3,6-anhydrogalactose is necessary to cause precipitation, and in the absence of the latter the λ-carrageenan remains in solution. The proportion of 3,6-anhydrogalactose in the molecule appears to be the major factor determining the concentration of potassium chloride necessary to bring about precipitation, but the amount of ester sulphate and perhaps its distribution also has its effect on the ease of precipitation (Pernas at $al.$, 1967).

The effect of caesium and rubidium is about the same as that of potassium in carrageenan gels (Bayley, 1955), but the former two ions have more influence than potassium on the setting temperature and strength of gels of the extract from $Hypnea$ $musciformis$ (DeLoach et $al.$, 1946).

It is of interest that highly substituted sodium cellulose sulphate (Schweiger, 1966), which is soluble in cold water, gives gels with potassium, rubidium and caesium ions.

It was found that the gel strength and melting point of carrageenan were increased by autoclaving with alkali (Marshall et $al.$, 1949), and a similar treatment is used in Japan to increase the gelling power of $Gracilaria$ $verrucosa$ extractives (Ushiyama and Koike, 1953). It has also been observed that alkali treatment of porphyran increased the viscosity and the tendency to gel out of solution in the presence of salts (Rees, 1961a). It is known (see pp. 138, 141) that this treatment eliminates 6-O-sulphate groups in carrageenan, with formation of 3,6-anhydro-residues, but whether the increased gel strength is a result of the reduction in sulphate ester content, or of the increase in 3,6-anhydrogalactose content is not known; probably both the changes have some effect. Other factors, not yet understood, must be involved in gel formation to explain why a number of galactans, for example funoran from $Gloiopeltis$ and the extract from $Laurentia$ $pinnatifida$, do not gel. Funoran contains about 31% 3,6-anhydro-L-galactose and 18·5% ester sulphate, and the galactan from $L.$ $pinnatafida$ is rich in both these constituents.

A. GEL STRENGTH

Many publications on seaweed extractives include figures for gel strength. When, as is sometimes the case, the strengths are given in arbitrary units, they clearly cannot be compared with those of other workers. Even when the figures are for the force in grams per square centimetre required to rupture the gel of a certain concentration, there is still scope for considerable variation in the results depending on the details of the method used, particularly the rate at which the force is applied (Selby and Selby, 1959; Goring and Young, 1955).

In addition to the uncertainty in comparing the results of different workers, there is the further complication that there can be a very wide range of gel strengths for extractives from one species of weed, so much so that it is possible only in extreme cases to say which species give stronger gels than others. This is hardly surprising as studies on individual species, for example *Chondrus crispus* by Black *et al.* (1965), have shown considerable variation in the proportions of the different galactans and in the proportions of 3,6-anhydrogalactose and of ester sulphate, two constituents which as mentioned above have an important influence on gelling behaviour.

Agar from *Gelidium* species probably gives the highest gel strengths consistently, but, in the presence of suitable concentrations of potassium salts, extracts from *Furcellaria* and those from *Chondrus crispus* with a high proportion of κ-carrageenan can give similar gel strengths to agar. On the other hand, extracts consisting largely of λ-carrageenan have little or no gel strength, and the same is true of κ-carrageenan in which the half sulphate ester is neutralized only with sodium.

B. VISCOSITY

The viscosity of solutions of the galactans forms part of the specification of many commercial grades, but it cannot be taken as an indication of molecular weight in view of the differences in chemical composition which can also affect viscosity. In general, agars give rather low viscosity solutions, below 10 cps in 1% solution at 50°, while 1% solutions of carrageenans have viscosities ranging from below 10 cps to over 1000 cps at 50°.

C. PRECIPITATION

The sensitivity of the galactans to different precipitating agents seems to depend largely on the proportions of 3,6-anhydro and sulphate ester groups, and as already described, fractionation can be achieved in some cases. Agar, but not carrageenan, is precipitated by tannic acid, while carrageenan combines, to give insoluble products, with a wide variety of cationic substances, among them methylene blue (Ewe, 1930; Graham, 1960) and cetylpyridinium chloride. The latter reagent was tested with a number of galactans (Graham and Thomas, 1962), and the amount bound to the hydrocolloid was found to vary considerably, being the least with agar and the greatest with λ-carrageenan.

VI. APPLICATIONS

As most of the red algae are small plants, the cost of collecting them is considerable and the galactans obtained from them are expensive compared with most industrial colloids. They are, therefore, used in cases where their properties are of particular value, or where they are effective in lower concentration than the cheaper materials.

A. AGAR AND AGAROSE

Most uses of agar depend on its gel-forming properties and it is used particularly when a gel with a high melting point is required. To comply with the specification of the U.S.A. Pharmacopeia, a 1·5% by weight solution of agar (prepared by boiling the solid with water for 10 min) must congeal at 32 to 39° to form a firm resilient gel which does not melt below 85°. The wide difference between setting and melting point, allowing inoculation at a temperature below that of incubation, is one reason for the almost universal use of agar in solidified bacteriological culture media. Other reasons are the stability at high temperatures, permitting sterilization without appreciable loss of gel strength, and the fact that very few microorganisms attack it.

Following the use of agar gels for the culture of microorganisms, their use has extended to the study of the metabolic products formed by these organisms and to other biochemical investigations. A common method of assay of antibiotics is the application of the material to an agar plate and measurement of the area free from growth of a test organism after incubation.

The different rates of diffusion of molecules of varying sizes through agar gels provides a method of purification of very high molecular weight substances such as viruses (Akers and Steere, 1961). Passing solutions through a column of particles of agar gel has proved to be the most effective method of using this property of the gels (Andrews, 1962). An agar gel is prepared and disintegrated with a rotating blade homogenizer and the particles of the desired size are separated by sieving. By using proteins of known molecular weights as standards, the molecular weights of other proteins in the range 16,000 to 670,000 can be determined.

In recent years agarose gels have proved more useful than those of unfractionated agar for some of these applications. For immunodiffusion and the electrophoretic separation of antigens (Brishammar et al., 1961) it has the advantage of giving a clear gel and because of its low sulphate content, it exhibits practically no absorption phenomena or electroendosmosis. In the gel diffusion analysis of basic antigens it gives more distinct lines of precipitation without halo effects.

The hardness of agarose gels of low concentrations enables gel particles of lower solids content to be used in columns for gel filtration, allowing the separation of larger molecules than is possible with unfractionated agar gel. The gel particles are prepared (Hjerten, 1964) by dispersing the sol in an immiscible organic liquid and keeping it stirred while it cools.

Considerable quantities of agar are used in canned meat, poultry and fish products as it will withstand the high temperature necessary for sterilization and the gel remains firm at temperatures at which gelatine gel would liquify.

Agar is also used as a bulk laxative both alone and in paraffin emulsions where it acts also as an emulsion stabilizer.

B. CARRAGEENAN-TYPE PRODUCTS

A very large number of different products extracted from red seaweeds are on the market. Species of *Chondrus* and *Gigartina* are most commonly used, but *Furcellaria*, *Euchuema*, *Hypnea* and some other species are also important.

They are prepared for particular applications, and specifications are based on performance in specially designed tests, rather than on chemical composition. It is probable, however, that those products which are designed for gel formation are rich in 3,6-anhydro-D-galactose, while those more suitable for giving viscosity and for suspending solids contain less 3,6-anhydro-D-galactose and more ester sulphate. Variations in the cations combined with the sulphate are also important in modifying behaviour.

Some important uses of carrageenan depend on its reactivity with proteins, particularly with the casein in milk. At the low concentration of 0·03% to 0·05%, a weak thixotropic gel is formed, and this property is widely used in the preparation of chocolate milk drinks where carrageenan keeps the cocoa powder suspended. Standardized preparations are on the market for use in preparing chocolate milk by either a hot or a cold process. The chemical differences making the products more suitable for one or the other type of preparation have not been published, but it has been reported (Gordon *et al.*, 1966) that extracts from *Iridaea laminarioides*, *Gigartina acicularis* and *G. pistillata* are effective when added to milk at 30°, but extracts from *Chondrus crispus* do not react with milk at temperatures below 50°.

Higher concentrations of carrageenan, particularly of types with a high content of the κ-fraction, give milk puddings with a jelly texture.

The strongly gelling types are used also as general gelling agents, particularly as with agar in canned products. Up to 0·1% of potassium chloride is generally included in the gel to improve its strength. Mixtures with some other colloids give gels with improved properties for food purposes, locust bean gum being particularly effective (Baker, 1949).

A small use, but one for which only weak carrageenan semi-gels give the required results, is in the "marbling" of paper for book covers. Pigments are sprinkled on to a crude extract of *Chondrus crispus* and distributed to give the traditional patterns to paper applied to this coloured layer.

Carrageenan is used in combination with other colloids in stabilizers for ice-creams and sherberts. Non-gelling types are used to control viscosity and suspend solids in toothpaste.

In recent years, promising medical uses have been found for λ-carrageenan. A degraded product appears to inhibit peptic activity and is of value in the treatment of peptic ulcers (Anderson and Soman, 1963). Probably as a result of the chemical similarity to animal mucopolysaccharides, it is effective in stimulating the growth of connective tissue in the guinea pig when injected subcutaneously (McCandless and Lehoczky-Mona, 1966).

Carrageenan has been used in the fractionation of agar (p. 129) to improve the precipitation of agaropectin by cationic surface active compounds (Blethen, 1966).

REFERENCES

Akers, G. K., and Steere, R. L. (1961). *Nature, Lond.* **192**, 436.
Anderson, N. S., and Rees, D. A. (1965). *J. chem. Soc.* p. 5880.
Anderson, N. S., and Rees, D. A. (1966). *Proc. 5th int. Seaweed Symp.* (1965), Halifax, Nova Scotia (E. G. Young and J. L. McLachlan, eds), p. 243, Pergamon Press, Oxford.
Anderson, W., and Soman, P. D. (1963). *Nature, Lond.* **199**, 389 and ref. cited therein.
Andrews, P. (1962). *Nature, Lond.* **196**, 36 and ref. cited therein.
Arai, K. (1961). *J. chem. Soc. Japan* **82**, 771, 1416.
Araki, C. (1937a). *J. chem. Soc. Japan* **58**, 1338.
Araki, C. (1937b). *J. chem. Soc. Japan* **58**, 1362.
Araki, C. (1940). *J. chem. Soc. Japan* **61**, 775.
Araki, C. (1956). *Mem. Fac. ind. Arts Kyoto tech. Univ.* (Sci. and Tech.) **5**, 21.
Araki, C. (1966). *Proc. 5th int. Seaweed Symp.* (1965), Halifax, Nova Scotia (E. G. Young and J. L. McLachlan eds), p. 3, Pergamon Press, Oxford.
Araki, C., and Arai, K. (1956). *Bull. chem. Soc. Japan* **29**, 339.
Araki, C., and Arai, K. (1957). *Bull. chem. Soc. Japan* **30**, 287
Araki, C., and Hirase, S. (1954). *Bull. chem. Soc. Japan* **27**, 105, 109.
Araki, C., and Hirase, S. (1956). *Bull. chem. Soc. Japan* **29**, 770.
Araki, C., and Hirase, S. (1960). *Bull. chem. Soc. Japan* **33**, 291, 597.
Baker, G. L. (1949). U.S. Patent, 2,466,146.
Barry, V. C., and McCormick, J. E. (1957). *J. chem. Soc.* p. 2777.
Bayley, S. T. (1955). *Biochim. biophys. Acta* **17**, 194.
Black, W. A. P., Blakemore, R. R., Colquhoun, J. A., and Dewar, E. T. (1965). *J. Sci. Fd. Agric.* **16**, 573.
Blethen, J. (1966). U.S. Patent, 3,281,409.
Bowker, D. M. (1966). Ph.D. Thesis, University of Wales.
Brishammar, S., Hjerten, S., and Hofsten, B. V. (1961). *Biochim. biophys. Acta* **53**, 518.
Buchanan, J., Percival, E. E., and Percival, E. G. V. (1943). *J. chem. Soc.* p. 51.
Christison, J., and Turvey, J. R. (1966). *Biochem. J.* **100**, 20P.

Clancy, M. J., Walsh, (Miss) K., Dillon, T., and O'Colla, P. S. (1960). *Proc. R. Dubl. Soc.* IA, 197.

Clingman, A. L., and Nunn, J. R. (1959). *J. chem. Soc.* p. 493.

Clingman, A. L., Nunn, J. R., and Stephen, A. M. (1957). *J. chem. Soc.* p. 197.

DeLoach, W. S., Wilton, O. C., Humm, H. J., and Wolf, F. A. (1946). *Duke Univ. Marine Sta. Bull.* No. **3**, 31.

Dewar, E. T., and Percival, E. G. V. (1947). *J. chem. Soc.* p. 1622.

Dillon, T., and McKenna, J. (1950a). *Nature, Lond.* **165**, 318.

Dillon, T., and McKenna, J. (1950b). *Proc. R. Ir. Acad.* **53**B, 45.

Dolan, T. C. S. (1965). Ph.D. Thesis, Edinburgh.

Dolan, T. C. S., and Rees, D. A. (1965). *J. chem. Soc.* p. 3534.

Ewe, G. E. (1930). *J. Am. pharm. Ass.* **19**, 568.

Forbes, I. A., and Percival, E. G. V. (1939). *J. chem. Soc.* p. 1844.

Gordon, A. L., Jonas, J. J., and Overholt, M. N. (1966). *Proc. 5th int. Seaweed Symp.* (1965), Halifax, Nova Scotia (E. G. Young and J. L. McLachlan, eds), p. 377, Pergamon Press, Oxford.

Goring, D. A. I., and Young, E. G. (1955). *Can. J. Chem.* **33**, 480.

Graham, H. D. (1960). *Fd. Res.* **25**, 720.

Graham, H. D., and Thomas, L. B. (1962). *Fd. Res.* **27**, 98.

Hands, S., and Peat, S. (1938). *Chemy. Ind.* p. 937.

Hassid, W. Z. (1933). *J. Am. chem. Soc.* **55**, 4163.

Hassid, W. Z. (1935). *J. Am. chem. Soc.* **57**, 2046.

Hegenauer, C., and Nace, G. W. (1965). *Biochim. biophys. Acta* **111**, 334.

Hirase, S. (1957). *Bull. chem. Soc. Japan* **30**, 68, 70, 75.

Hirase, S., and Araki, C. (1961). *Bull. chem. Soc. Japan* **34**, 1048.

Hirase, S., Araki, C., and Ito, T. (1956). *Bull. chem. Soc. Japan* **29**, 985.

Hirase, S., Araki, C., and Ito, T. (1958). *Bull. chem. Soc. Japan* **31**, 428.

Hjerten, S. (1961). *Biochim. biophys. Acta* **53**, 514.

Hjerten, S. (1962). *Biochim. biophys. Acta* **62**, 445.

Hjerten, S. (1964). *Biochim. biophys. Acta* **79**, 393.

Hoppe, H. A. (1966). *Botanica mar.* **9**, 5.

Johnston, R., and Percival, E. G. V. (1950). *J. chem. Soc.* p. 1994.

Lipatov, S. M., and Morozov, A. A. (1935). *Kolloidzeitschrift* **71**, 317.

Lund, S., and Bjerre Petersen, E. (1964). *Proc. 4th int. Seaweed Symp.* (1961), Biarritz, (A. D. Devirville and J. Feldmann, eds), p. 410, Pergamon Press, Oxford.

McCandless, E. L., and Lehoczky-Mona, J. (1966). *Proc. 5th int. Seaweed Symp.* (1965), Halifax, Nova Scotia (E. G. Young and J. L. McLachlan, eds), p. 157, Pergamon Press, Oxford.

Marshall, S. M., Newton, L., and Orr, A. P. (1949). "A Study of Certain British Seaweeds and their Utilization in the Preparation of Agar", H.M.S.O., London.

Morgan, K. and O'Neill, A. N. (1959). *Can. J. Chem.* **37**, 1201.

Mori, T. (1950). *Chem. Abstr.* **44**, 7783.

Morozov, A. A. (1935). *Colloid J.* (U.S.S.R.) **1**, 37.

Nunn, J. R., and von Holdt, M. M. (1957). *J. chem. Soc.* p. 1094.

O'Colla, P. S., MacCraith, D., and NiOlain, R. M. (1958). *Abs. 3rd int. Seaweed Symp.*, Galway, Ireland, p. 77.

O'Colla, P. S. (1962). *In* "Physiology and Biochemistry of Algae" (R. Lewin, ed.), Chapter 20, Academic Press, New York and London.

Ohmi, H. (1958). *Mem. Fac. Fish. Hokkaido Univ.* **6**, 1.

O'Neill, A. N. (1955a). *J. Am. chem. Soc.* **77**, 2837.

O'Neill, A. N. (1955b). *J. Am. chem. Soc.* **77**, 6324.

O'Neill, A. N., and Stewart, D. K. R. (1956). *Can. J. Chem.* **34**, 1700.

Painter, T. J. (1960). *Can. J. Chem.* **38**, 112.

Painter, T. J. (1967). *Proc. 5th int. Seaweed Symp.* (1965), Halifax, Nova Scotia (E. G. Young and J. L. McLachlan, eds), p. 305, Pergamon Press, Oxford.

Peat, S., and Turvey, J. R. (1965). *Fortschr. Chem. org. Natstoffe* **23**, 1.

Peat, S., Turvey, J. R., and Rees, D. A. (1961). *J. chem. Soc.* p. 1590.

Percival, Elizabeth (1954). *Chemy. Ind.* p. 1487.

Percival, E. G. V., and Somerville, J. C. (1937). *J. chem. Soc.* p. 1615.

Pernas, A. J., Smidsrod, O., Larsen, B., and Haug, A. (1967). *Acta chem. scand.* **21**

Rees, D. A. (1961a). *J. chem. Soc.* p. 5168 and ref. cited therein.

Rees, D. A. (1961b). *Biochem. J.* **81**, 347.

Rees, D. A. (1963). *J. chem. Soc.* p. 1821.

Rees, D. A. (1965). "Carbohydrate Sulphates" in Ann. Reports, p. 469.

Rees, D. A., and Conway, E. (1962a). *Biochem. J.* **84**, 411.

Rees, D. A., and Conway, E. (1962b). *Nature, Lond.* **195**, 398.

Russell, B., Mead, T. H., and Polson, A. (1964). *Biochim. biophys. Acta* **86**, 169.

Schweiger, R. G. (1966). *Chemy. Ind.* 900.

Selby, H. H., and Selby, T. A. (1959). *In* "Industrial Gums" (R. L. Whistler and J. N. BeMiller, eds), p. 15, Academic Press, New York and London.

Smith, D. B., and Cook, W. H. (1953). *Archs. no. biochem. Biophys.* **45**, 232.

Smith, D. B., Cook, W. H., and Neal, J. L. (1954). *Archs. no. biochem. Biophys.* **53**, 192.

Smith, D. B., O'Neill, A. N., and Perlin, A. S. (1955). *Can. J. Chem.* **33**, 1352 and ref. cited therein.

Stoloff, L. (1959). *In* "Industrial Gums" (R. L. Whistler and J. N. BeMiller, eds), p. 91, Academic Press, New York and London.

Stoloff, L., and Silva, P. (1957). *Econ. Bot.* **11**, 327.

Su, J. C. (1958). *Nature, Lond.* **182**, 1779.

Su, J. C., and Hassid, W. Z. (1962). *Biochemistry* **1**, 468.

Turvey, J. R., and Williams, T. P. (1961). *Colloques. int. Cent. natn. Rech. scient.* **103**, 29.

Turvey, J. R., and Rees, D. A. (1961). *Nature, Lond.* **189**, 831.

Turvey, J. R., and Williams, T. P. (1964). *Proc. 4th int. Seaweed Symp.* (1961), Biarritz (A. D. DeVirville and J. Feldmann, eds), p. 370. Pergamon Press, Oxford.

Turvey, J. R., and Simpson, P. R. (1966). *Proc. 5th int. Seaweed Symp.* (1965), Halifax, Nova Scotia (E. G. Young and J. L. McLachlan, eds), p. 323, Pergamon Press, Oxford.

Ushiyawa, S., and Koike, H. (1953). Jap. Pat. 2286 (see *Chem. Abstr.* 1954 **48**, 6619).

Weigl, J., and Yaphe, W. (1966). *Can. J. Microbiol.* **12**, 939.

Weigl, J., Turvey, J. R., and Yaphe, W. (1966). *Proc. 5th int. Seaweed Symp.* (1965), Halifax, Nova Scotia (E. G. Young and J. L. McLachlan, eds), p. 329, Pergamon Press, Oxford.

Wu, Y. C., and Ho, H. K. (1959). *J. Chinese Chem. Soc.* **6**, 84.

Yaphe, W. (1957). *Can. J. Microbiol.* **3**, 987.

Yaphe, W. (1959). *Can. J. Bot.* **37**, 751.

Yaphe, W. (1966). *Proc. 5th int. Seaweed Symp.* (1965), Halifax, Nova Scotia (E. G. Young and J. L. McLachlan, eds), p. 333, Pergamon Press, Oxford.

CHAPTER 7

Sulphated Polysaccharides Containing Neutral Sugars: 2

I. FUCOIDAN

Fucoidin, first isolated and named by Kylin (1913, 1915), is more systematically named fucoidan. It is thought to occur in the intercellular tissues or mucilaginous matrix (see p. 12) (Black, 1954; McCulley, 1965) and it has been shown to be present in the droplets which exude from the surface of the frond of *L. digitata* (Lunde *et al.*, 1937), *A. nodosum* (Dillon *et al.*, 1953) and *Macrocystis* (Schweiger, 1962a), although this exudate also contains alginic acid (Black, 1954; E. Percival, unpublished work). It is very hygroscopic and it may therefore serve to prevent dehydration of the plant upon long exposure which would explain the higher fucoidan content of plants at intertidal level (see p. 9).

Kylin isolated fucoidan from *L. digitata*, *Fucus vesiculosus* and *Ascophyllum nodosum* and, after hydrolysis, separated the fucose as the phenyl-L-fucosazone and claimed that pentoses were also present in the hydrolysate. Subsequent studies on fucoidan separated by hot water extraction from *F. vesiculosus*, *F. spiralis*, *L. hyperborea* and *Himanthalia lorea* (Percival and Ross, 1950) revealed the presence of half ester sulphate (38%) and only 56·7% of fucose even in the most highly purified sample (from *H. lorea*) which also contained 4% galactose, 1·5% xylose, about 3% uronic acid and 8% of metals. A calcium fucan monosulphate $(C_6H_9O_3SO_4Ca0·5)n$ would give sulphate 39%, calcium 8% and fucose 66·9%. Other workers (Dillon *et al.*, 1953) separated from the seed mucilage of *A. nodosum* a sulphated polysaccharide containing fucose to galactose in the ratio of 8:1. Fractionation of a sample of fucoidan from *F. vesiculosus* on DEAE-cellulose gave a small quantity of a xylan (see p. 88) together with several fractions which contained fucose and small proportions of galactose. No fraction comprising only fucose residues was separated (Lloyd, 1960). Schweiger (1962b) examined the exudate from *Macrocystis pyrifera* and, after extensive purification, isolated a galactofucan in which the ratio of fucose to galactose remained constant at 18:1 during a variety of purification procedures, although the amount of xylose varied from 0·5 to 2·5%. This author considers that the existence of a pure fucan sulphate as a major constituent in this mucilage is unlikely, but that in addition to the galactofucan the presence of a small quantity of a xylogalactofucan is possible. At the same time it should be pointed out that purification of a crude fucan preparation from *F. vesiculosus* by fractionation with ethanol containing 0·3% sodium acetate and further purification by fractional precipitation with ethanol and acetone and with cetyl dimethylbenzylammonium chloride (Bernardi and Springer, 1962) gave two fractions which had fucose contents of 63 and 65%, ester sulphate about 30%, ash about 29% and $[\alpha]_D$ $-137°$ and $-140°$, respectively. They gave on hydrolysis fucose and trace quantities of xylose and an unidentified fast sugar (with a chromatographic mobility slightly different from 3-O-methylfucose). These fractions were nitrogen-free and uronic acid-free and no significant heterogeneity could be demonstrated either physically or chemically.

Evidence for more than a single molecular species in fucoidan has also been revealed by free-boundary electrophoresis by Larsen and Haug (1963) on material extracted from *A. nodosum* and from *F. vesiculosus* by O'Neill (1954). In citrate buffer at pH 4·2, three boundaries were observed for the *A. nodosum* extract and only two boundaries with mobilities of 2·3 and 2·5 × 10^{-4} cm^2 per sec volt for the *F. vesiculosus* material (acetate buffer at pH 5·0). The separation of three components from the former was somewhat better at pH 2·0 (see Fig. 1).

→+ ascending boundaries   →+ ascending boundaries
1% Solutions in 0·05M-citrate buffer   1% Solutions at pH 2

FIG. 1.

While these results may merely indicate the presence of similar polysaccharides with different degrees of sulphation, it is possible that they represent polysaccharides of different composition.

The situation has recently become more confused by the separation from *A. nodosum* of at least three distinct fucose containing polysaccharides (Larsen *et al.*, 1966; Percival, 1967) all of which contain varying proportions of fucose, xylose, glucuronic acid, ester sulphate and protein (see p. 176). Since these materials defined separation into polysaccharides in which the uronic acid was less than 5%, they will, for convenience, be dealt with in Chapter 8.

A. ISOLATION

If a fucoidan having mucilaginous properties is desired, the methods of isolation given below must be modified since complete drying is detrimental to the colloidal properties of fucoidan and fresh wet weed should be employed as raw material (Lunde *et al.*, 1937). Precipitation with ethanol will yield material with mucilaginous properties but it is invariably contaminated with alginic acid and protein, and attempts to separate such impurities usually destroy the colloidal properties.

Fucoidan can be extracted with water or acid from the dried weed.

1. *Purification as the Lead Hydroxide Complex*

The dried ground weed (100 g) is extracted with water (300 ml) on a boiling water-bath for 24 hr (Percival and Ross, 1950). After filtration through muslin, the extract is treated with lead acetate until precipitation of alginates and protein is complete. Addition of barium hydroxide to the filtrate until the solution is just alkaline to phenolphthalein causes the separation of a fucoidan–lead hydroxide complex. Decomposition of the complex is achieved by shaking overnight with sulphuric acid (30 ml 4N-sulphuric acid in 100 ml water), and the resulting solution is freed from a considerable amount of colouring matter by filtration through Filter Cel. The crude fucoidan (3·8 g from *Himanthalia lorea*) is isolated, after dialysis and concentration, by precipitation with alcohol. This is purified by dissolution in water and two treatments with charcoal (1%) and Filter-Cel (2%) at 90° for 30 min. The fucoidan (2·3 g) isolated as a white powder has $[\alpha]_D -140°$ (conc. 1·0 in water) and contains approximately 44% fucose, 32·5% sulphate, 22·5% ash and 7% metals. It is exceedingly difficult to free from moisture. After drying a sample at 40° and 0·1 mm pressure for 18 hr it still retained 9·4% water and 6% ethanol.

2. *Purification by Alcohol Precipitation*

The dried milled seaweed is stirred with ten parts (w/v) hydrochloric acid at pH 2 to 4·5 at 100° for 3 to 7½ hr (Black *et al.*, 1952). This extracts 50 to 60% of the fucoidan and two further treatments extracts up to 80%. Neutralization of the extracts and evaporation to dryness yields the crude fucoidan as a brown residue. Partial purification is achieved by dissolution in water and fractional precipitation with ethanol at 30 and 60% concentration. Impurities are precipitated at 30% concentration and fucoidan is deposited at 60% concentration.

Treatment of an aqueous solution of this partly purified material with formaldehyde and evaporation *in vacuo* yields a glassy solid. The fucoidan can be extracted from this with water while most of the impurities remain insoluble. The fucoidan is isolated from the aqueous solution by precipitation with ethanol (70% concentration) and dried with ethanol and ether (loss in purification 20 to 30%).

B. DETERMINATION OF THE FUCOIDAN CONTENT OF THE WEED

Since analysis of fucoidan by actual isolation of the material is tedious, the more convenient determination of the fucose content of the weed is usually employed, and these results can be multiplied by 1·75 to convert them into the approximate fucoidan content.

1. Boiling the seaweed with hydrochloric acid liberates the fucose from fucoidan and converts it into 5-methyl-2-furfuraldehyde which is distilled into a solution of phloroglucinol. The precipitated phloroglucide is weighed.

2. Hydrolysis to fucose with 2·5% sulphuric acid at 100° for 3 hr is followed by filtration of the residual weed. The amount of fucose in the filtrate can be determined:

(a) gravimetrically as the insoluble methylphenylhydrazone (yield 93·4%) (Percival, 1962).

(b) by oxidation with periodate whereby the terminal —CHOH—CH_3 group is converted into acetaldehyde. This is then estimated by: (i) a modification of the method of Nicolet and Shinn (Cameron et al., 1948). The acetaldehyde is passed into a solution of sodium bisulphite with which it forms a complex. After the excess bisulphite has been oxidized with iodine, the complex is decomposed with a saturated solution of bicarbonate and the liberated bisulphite is titrated with iodine. This titration is slow, the iodine being added drop by drop until a permanent starch-iodine blue colour remains for 2 min. Further bicarbonate solution is then added to ensure that all the aldehyde-bound bisulphite has been liberated. (ii) The acetaldehyde may also be determined colorimetrically by the method of Fromageot and Heitz (1938) (see Black et al., 1950) which depends on the blue colour produced by acetaldehyde with piperazine and sodium nitroprusside.

C. CONSTITUTION

Although fucoidans isolated from different genera of brown seaweeds appear to have somewhat different compositions, they all, as far as is known, with the exception of some of the molecular species synthesized by A. nodosum, comprise mainly fucose and ester sulphate (Quillet, 1961).

In an investigation of a sample of fucoidan from F. vesiculosus (fucose 38%, sulphate 32·8%) $[\alpha]_D$ −115° (Conchie and Percival, 1950), it was found that about 10% of the total ester sulphate was labile to alkali, indicating that a small proportion of the sulphate groups are linked to C-2 or C-3 of 1,4-linked fucose residues (vicinal *trans* hydroxyl groups are necessary for alkali lability) (Percival, 1949). At the same time it follows for a linear molecule that the main linkage of the L-fucopyranose units cannot be through the hydroxyl on C-4, since this would necessitate sulphate groups on C-2 or C-3 both of which would be alkali liabile.

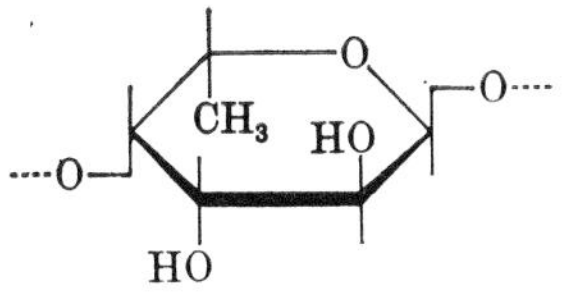

α-1,4-linked
L-fucopyranose unit

The main hydrolysis product (three-fifths of the total) from this sample of fucoidan after methylation was 3-*O*-methyl-L-fucose (**4**) (Conchie and Percival, 1950). Because of the alkali stability of the majority of the sulphate, such residues in the native polymer must be 1,2-linked and carry sulphate on C-4 [Fig. 2 (**1**)]. The strongly negative rotations of fucoidan and its derivatives provide evidence that these linkages are predominantly of the α-type.

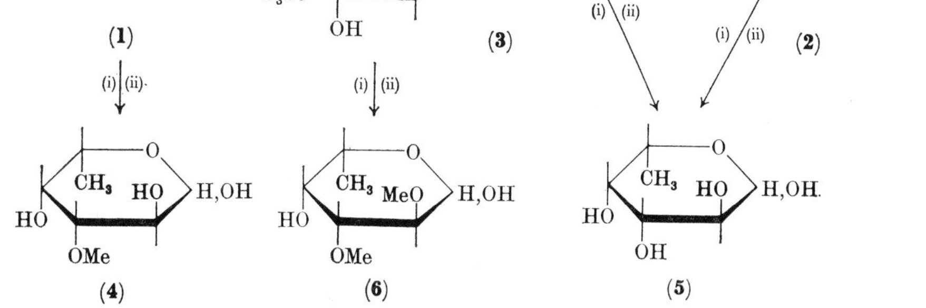

(i) = Methylation,
(ii) = Hydrolysis

FIG. 2.

The remaining sugars isolated after methylation and hydrolysis were free fucose (5) (one-fifth) and 2,3-di-O-methylfucose (6) (one-fifth). Excluding the possibility of incomplete methylation, the former could arise from a di-sulphated residue (2) or it could be derived from a branch point carrying a terminal group having free hydroxyls on C-2 and C-3 (3); the latter yielding the 2,3-dimethylsugar (6).

From the proportions of the different derivatives isolated it can be cal-culated that there is an average of one such branch in every five fucose units. The following two formulae (Fig. 3) conform with these requirements.

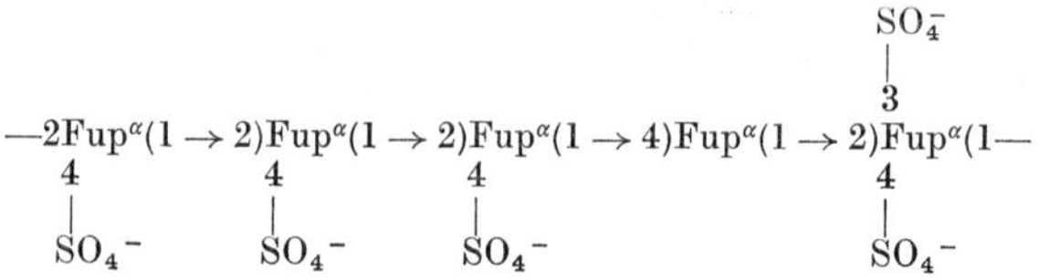

$$\begin{array}{c} \mathrm{SO_4^-} \\ | \\ 3 \\ \mathrm{-2Fup^\alpha(1 \to 2)Fup^\alpha(1 \to 2)Fup^\alpha(1 \to 4)Fup^\alpha(1 \to 2)Fup^\alpha(1-} \\ 4 \qquad\qquad 4 \qquad\qquad 4 \qquad\qquad\qquad\qquad 4 \\ | \qquad\qquad | \qquad\qquad | \qquad\qquad\qquad\qquad | \\ \mathrm{SO_4^-} \qquad \mathrm{SO_4^-} \qquad \mathrm{SO_4^-} \qquad\qquad\qquad \mathrm{SO_4^-} \end{array}$$

(7)

$$\begin{array}{c} \qquad\qquad\qquad\qquad (3 \to 1(\mathrm{Fup4SO_4^-} \\ \qquad\qquad\qquad\qquad \diagup \\ \mathrm{-2)Fup^\alpha(1 \to 2)Fup^\alpha(1 \to 2)Fup^\alpha(1 \to 2)Fup^\alpha(1-} \\ 4 \qquad\qquad 4 \qquad\qquad 4 \qquad\qquad 4 \\ | \qquad\qquad | \qquad\qquad | \qquad\qquad | \\ \mathrm{SO_4^-} \qquad \mathrm{SO_4^-} \qquad \mathrm{SO_4^-} \qquad \mathrm{SO_4^-} \end{array}$$

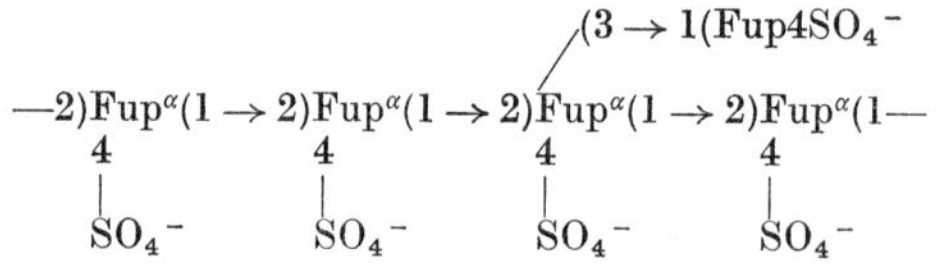

(8)

Fup = L-fucopyranose

FIG. 3.

Acetolysis studies by O'Neill (1954) and by Côté (1959) confirmed these findings. The former, after deacetylation, reduced the derived sugars to sugar alcohols, reacetylated and fractionated the product by column chromato-graphy. He finally isolated and characterized crystalline 2-O-α-L-fucopyrano-syl-L-fucitol (9).

Employing a commercial sample of fucoidan $[\alpha]_\mathrm{D}$ $-93°$ from *F. vesiculosus*, Côté separated not only the 1,2-linked disaccharide (10), but also the 1,4- (11) and a small quantity of the 1,3-linked disaccharides (12). The last could be derived from the branch point shown in Fig. 2 (8).

Evidence as to whether the galactose and xylose form part of the macro-molecule of fucoidan or are contaminating polysaccharides has not been obtained.

Little is known of the molecular size of this polymer. Osmometric deter-minations gave molecular weight values of 133,000 $\pm$ 20,000 (O'Neill, 1954), and determinations of a highly purified sample of fucoidan by means of the ultracentrifuge and Svedberg's formula gave a figure of $7 \cdot 8 \times 10^4$, but this latter figure must be accepted with reserve because of the incertitude in the value of the partial specific volume (Bernardi and Springer, 1962). In salt

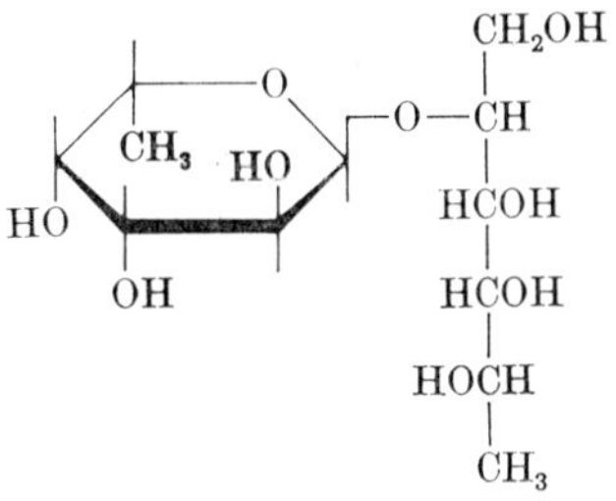

(9)

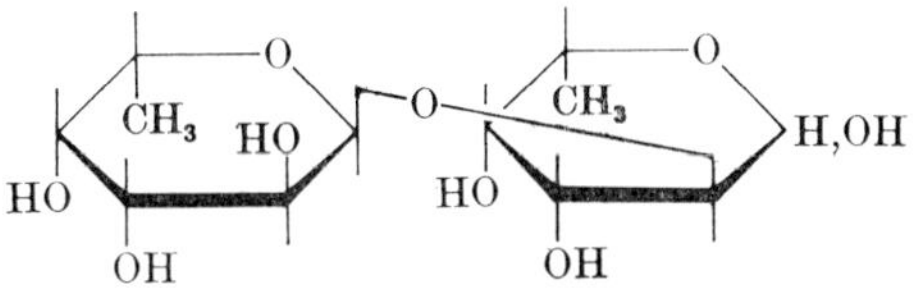

(10)

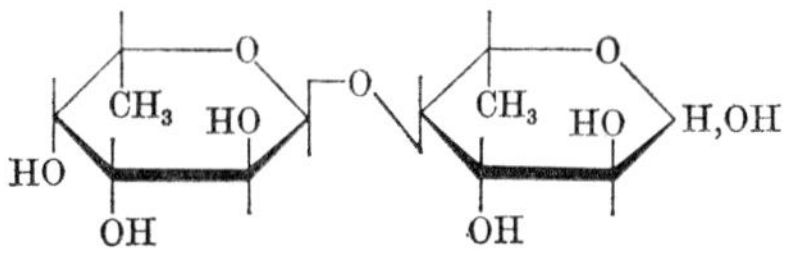

(11)

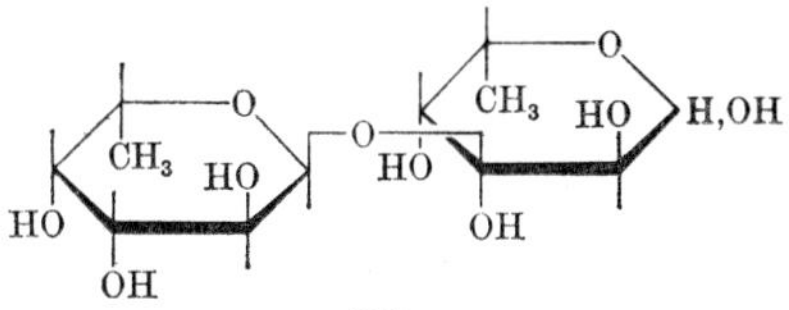

(12)

solution the fucan molecule is considered to be a highly-charged, branched, random coil. Its electrophoretic mobility -27×10^{-5} cm^2/sec/v at pH 5 which is much higher than that reported for the carrageenans is difficult to explain.

D. PROPERTIES

When freshly prepared by alcohol precipitation, fucoidan is a tan to off-white powder. It is soluble in water and insoluble in organic solvents. Specific rotations as far apart as $-75°$ (Nelson and Cretcher, 1931) and $-140°$ (Percival and Ross, 1950) have been reported. The purified fractions prepared by Bernardi and Springer were readily dissolved as 3% (w/v) in water, in physiological saline (0·85%) and all buffers tested. Aqueous solutions 1% of suitably isolated fucoidan (Lunde et al., 1937) are viscous and exhibit dilatent properties with pronounced mucilaginous or stringy-type of flow (McNeely, 1959). On standing for days or weeks, the viscosity drops and some precipitation occurs. The effect of adjusting the pH below 4 or above 9, dialysis, repeated precipitation from water, or excessive dehydration, is in all cases to impair or destroy the viscosity. Solutions of fucoidan in 65 to 70% sucrose syrup show pronounced mucilaginous properties, even in low concentration, and it is possible to pull out threads several feet long from them. Such solutions are very stable and retain their properties for years. As solutions of fucoidan in high concentrations of glycerol and other polyhydric materials exhibit similar properties to a lesser degree, it is suggested (McNeely, 1959) that the presence of a high concentration of polyhydric substance leads to the formation of aggregates through hydrogen bonding thus reducing the solubility compared with that in water. The effect of excessive drying might be to carry the aggregation a stage further leading to an insoluble product.

Dilute acids or bases cause irreversible changes in properties. The product obtained by acid treatment of the seaweed (Black *et al.*, 1952) is soluble in water in high concentrations as the calcium, potassium or aluminium salts; the solutions are of low viscosity and have no mucilaginous properties, indicating that considerable degradation has occurred.

E. ENZYMIC HYDROLYSIS

It was found that 31·5% and 29·9% of fucoidan was utilized, respectively, by *Pseudomonas atlantica* and *P. carrageenovora*, when these two marine bacteria were grown on artificial media in which fucoidan was the sole carbon source (Yaphe and Morgan, 1959). The cell-free medium was then used as the enzyme solution and tested against fucoidan. The enzyme activity was highest at about pH 7, and after 24 hr at 40° and this pH, the apparent degree of conversion of fucoidan to fucose was 24·2% by preparations from *P. atlantica* and 15·9% from *P. carrageenovora*. After heating the *P. atlantica* preparation at 100° for 10 min it still showed 8·1% hydrolysis of fucoidan. A similar enzyme was demonstrated in the mid gut of the snail *Tegula* (Galli and Giese, 1959). Enzymic extracts from *Rhodymenia palmata*, *Cladophora rupestris* and *Ulva lactuca* were devoid of hydrolytic activity towards fucoidan (Duncan *et al.*, 1956).

F. USES

Fucoidan has been suggested as a source of the rare sugar L-fucose (Black *et al.*, 1953; Schweiger, 1962a). It has also been tested as a blood anticoagulant. Fractions with blood anticoagulant activity have been isolated from crude fucoidan in 25% yield (Springer *et al.*, 1957) and the highly purified fractions described on page 158 had 60 to 80% of the activity of heparin in the recalcification test, and 15 to 18% of the heparin activity in the whole human blood coagulation inhibition assay. Furthermore, these fractions were considerably more effective than heparin in delaying the action of thrombin on fibrinogen. Certain fucoidan fractions, although less active anticoagulants, were potent lipaemia clearers (Schuler and Springer, 1957).

Ascophyllum nodosum on breakwater, Forth
Bridge, Scotland.

CLADOPHORA rupestris. Kütz.

Nature Printed by Henry Bradbury

II. SULPHATED WATER-SOLUBLE POLYSACCHARIDES OF THE CHLOROPHYCEAE

Many of the Chlorophyceae synthesize sulphated polysaccharides containing a wide variety of sugars, and although partial fractionation has been achieved in some instances, in no case has a sulphated homopolysaccharide been separated.

A. CLADOPHORA RUPESTRIS

The first polysaccharide comprising only neutral sugars and ester sulphate to be investigated chemically (Fisher and Percival, 1957) was the water-soluble material from *Cladophora rupestris*. The original extract contained about 25% of protein and all attempts to reduce this below about 8% were unsuccessful. The material finally investigated had $[\alpha]_D$ +69° and contained L-arabinose (3·7), D-galactose (2·8), D-xylose (1·0), L-rhamnose (0·4) and D-glucose (0·2) (approximate molar proportions in brackets), nitrogen 1·26%, ester sulphate 19·6% and ash 13·7%.

Extraction of the acetylated polysaccharide with chloroform removed all the glucose as a glucose-rich fraction devoid of sulphate. After removal of the acetate groups, this fraction gave no colour with iodine and reduced one mole of periodate for every 420 g glucan, indicating the apparent absence of starch and the presence of some 1,3-linked residues. Later work (Love *et al.*, 1963) (see p. 79) revealed the presence of about 1% of starch in a water-soluble extract which had been isolated under very mild conditions. It is very probable that the more drastic methods of extraction and purification had destroyed the starch in the above experiments.

Periodate oxidation of the material with $[\alpha]_D$ +69° reduced one molecule of periodate for every 347 g of polysaccharide, and the oxopolysaccharide (recovered in 80 to 87% yield) contained 20% of sulphate, 1·3% of nitrogen and the following proportions of uncleaved sugars: arabinose, 61·4%; galactose, 16·6%; rhamnose about 12% and glucose about 7%. About two-thirds of the galactose and all the xylose had been cleaved by this treatment and these residues must therefore contain free glycol groupings. In other words, the xylose must be present as end-group [Fig. 4 (**13**)] and/or as 1,4-linked units (**14**) and the galactose as 1,2, 1,4 or 1,6-linked (**15**) or end-group galactose (**16**). The presence of 2,3,4-tri- and 2,3-di-*O*-methylxylose in the hydrolysate of the methylated material, and of 2,3,4,6-tetra-, 2,3,5- and 2,3,4-tri-*O*-methyl-galactose confirmed the oxidation results and assigned end group and 1,6-linkage to those galactose units which were cleaved by the periodate.

The 2,4-di-*O*-methyl, 2-*O*-methyl and free galactose also present in the methylated hydrolysate could arise from 1,3-linked galactose units carrying

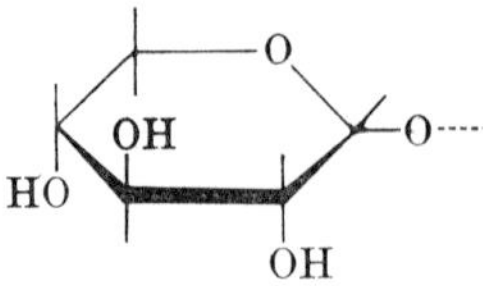

(13) 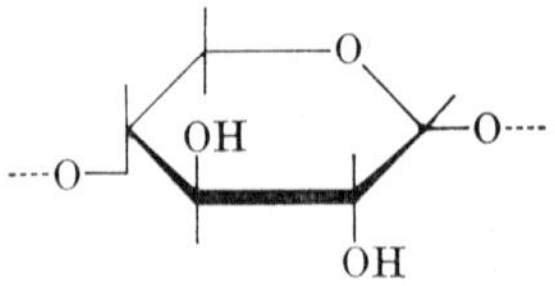   (14)

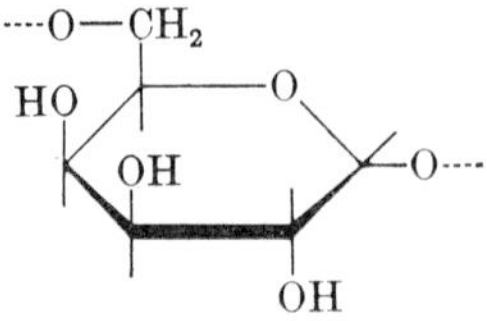

(15) 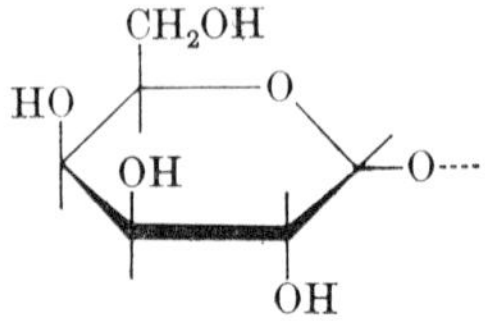   (16)

FIG. 4.

sulphate on the other free hydroxyl groups. Confirmation of this was obtained by the separation of β-(1 → 3)- [Fig. 5 (**17**)] and β-(1 → 6)- (**18**) galactobioses and of galactose 6-sulphate (**19**) from a partial acid hydrolysate of a subsequent extract from *C. rupestris* (Hirst *et al.*, 1965).

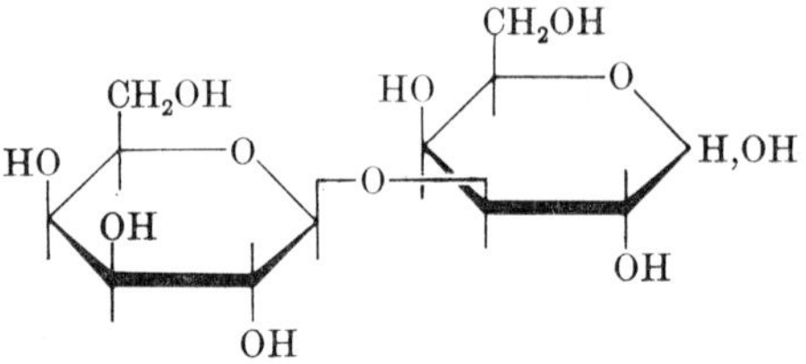

(17)

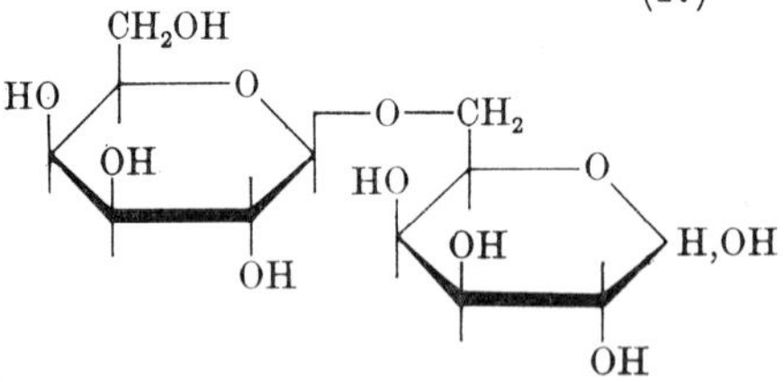

(18) 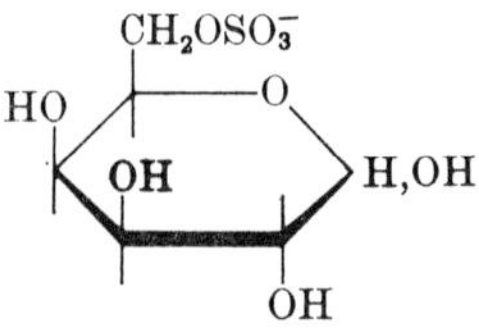   (19)

FIG. 5.

Some 2,4-di-*O*-methyl-L-arabinose was separated from the methylated hydrolysate but by far the largest fractions consisted of 2-*O*-methyl- and unmethylated L-arabinose indicating that this sugar was mainly linked or sulphated on C-3, C-3 and C-4. Such units would be immune to periodate and this explains the presence of a large proportion of uncleaved arabinose in the oxopolysaccharide. Partial hydrolysis studies support these results; arabinose 3-sulphate [Fig. 6 (20)] was separated and characterized and a 3-sulphated $\beta(1 \rightarrow 4/5)$ L-arabinose (21) or (22) was also isolated. The small amount of 2,4-di-O-methylarabinose was probably derived from 3-sulphated end-group arabinose.

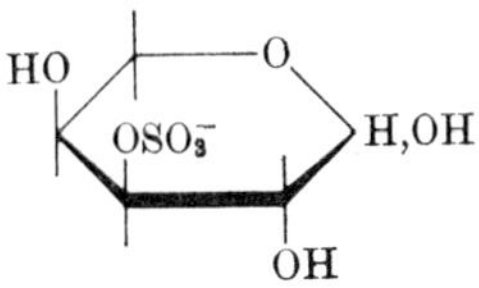

FIG. 6.

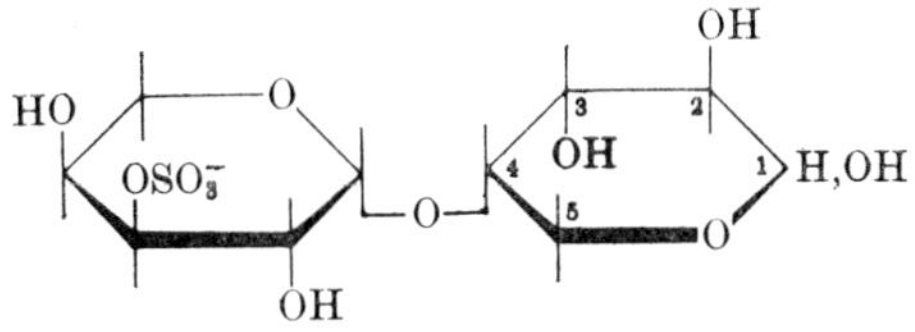

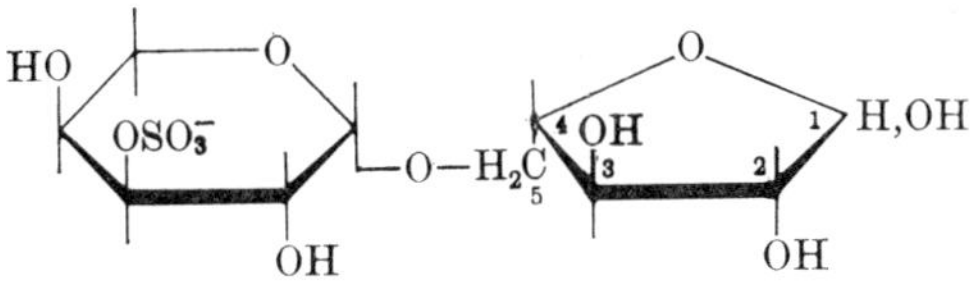

The comparatively large amount of free galactose and arabinose in the methylated hydrolysate no doubt arose in part from under methylation, such highly charged molecules being very difficult to methylate completely.

Confirmation of a number of these linkages was obtained from methylation of the polysaccharide after two Barry degradations (see p. 41) and a third oxidation with periodate (O'Donnell and Percival, 1959).

The failure to detect 3,6-anhydrogalactose after treatment of the polysaccharide with alkali (see p. 46) indicates that the galactose 6-sulphate must be linked at C-3.

Summarizing, the major structural units detected in this polysaccharide are $(SO = -SO_3^-)$:

$$Xyl(1 \rightarrow \quad ; \quad Gal(1 \rightarrow \quad ; \quad \rightarrow 4)Xyl(1 \rightarrow \quad ; \quad Gal(1 \rightarrow 6)Gal \quad ;$$

$$\overset{\displaystyle 6S}{\underset{\displaystyle |}{Gal(1 \rightarrow 3)}}\overset{\displaystyle 3S}{\underset{\displaystyle |}{Gal}} \quad ; \quad Ara(1 \rightarrow 4/5)Ara \quad ; \quad \overset{\displaystyle 3S}{\underset{\displaystyle |}{\rightarrow 4/5)}}\underset{\displaystyle \underset{\displaystyle |}{2}}{Ara(1 \rightarrow}$$

Although the balance of evidence indicates that this is a single heteropolysaccharide, all attempts to obtain unequivocal proof by the isolation of hetero-oligosaccharides from fragmentation studies have been unsuccessful. At the same time, the application of Barry degradation (see p. 41) supports the evidence for a single heteropolysaccharide since after three oxidations and two degradations, a 25% yield of degraded polymer was isolated which still contained 12·5% of sulphate (as SO_3^{2-}), arabinose, galactose and rhamnose in the molar proportions of 1:1:0·56 and was sufficiently large to be retained by a dialysis sac. It also provides further evidence of the highly branched nature of this polysaccharide which has xylose and galactose units at the periphery of the molecule and galactose, arabinose and rhamnose in the inner core. Sulphate groups are linked to residues both in the outer branches and in the centre of the molecule. Apart from adjacent galactose and arabinose units deduced from the isolation of the above disaccharides, no evidence is available on the relative positions of the sugars in the macromolecule.

In a purified sample of the *Cladophora* polysaccharide, extracted under mild conditions and from which the starch had been removed as the iodine complex, there was still 2·5% of glucose. Of this, 32% was degraded by salivary α-amylase and was presumably starch which had escaped precipitation; of the rest, 20% was not cleaved by periodate, as would be the case of the β-1,3-linked polymer separated as the acetate (see p. 165), and the remaining 48% was cleaved by periodate and not attached by α-amylase and could therefore be a 1,4-linked β-glucan. These glucans are apparently present as separate neutral polysaccharides and it is surprising that they were not removed when fractionation of this material on DEAE-cellulose was investigated (Hirst *et al.*, 1965).

B. *CHAETOMORPHA* SPP.

Parallel studies on *Chaetomorpha linum* and *Ch. capillaris* water-soluble polysaccharides (Hirst *et al.*, 1965) indicated the essential similarity of these materials to the corresponding polysaccharide of *C. rupestris*. The former two polysaccharides, after removal of an amylopectin (see p. 79) had similar rotations, sulphate, ash and protein contents and contained the same sugars as the *Cladophora* material, Tables 1 and 2. The proportions of galactose, xylose and protein were somewhat lower in the *Chaetomorpha* extracts but this can partly be explained by the fact that the solutions had been treated with trichloracetic acid before isolation of the polysaccharides.

Even so, the *Chaetomorpha* polysaccharides appear to have a galactose content which is lower than that of acid treated *Cladophora* polysaccharide.

Owing to the difficulty of getting large supplies of the *Chaetomorpha* species, it was not possible to carry out major structural studies on the extracts from these weeds, but similar qualitative experiments were made wherever possible. Partial acid hydrolyses of the *Chaetomorpha* polysaccharides gave the same chromatographic and ionophoretic pattern of free sugars, sulphated fragments and oligosaccharides as the *Cladophora* material and the same was

Table 1.

Polysaccharide from:	Sugar%	$[\alpha]_D$	Sulphate%	Ash%	Protein%
C. rupestris	43·3	+53°	12·3	8·1	25·6
Ch. capillaris	43·5	66	15·2	9·2	19·4
Ch. linum	43·5	75	15·5	9·3	18·8

Table 2.

Polysaccharide from:	Molar proportions of sugars		
	Galactose	Arabinose	Xylose
Cladophora rupestris[a]	3·1	3·2	1·0
Cladophora rupestris[b]	2·8	3·7	1·0
Ch. capillaris[b]	1·4	3·7	1·0
Ch. linum	1·8	3·7	1·0

[a] Extracted under mild conditions.   [b] Treated trichloroacetic acid.

true of the hydrolysate obtained after sodium methoxide treatment (see p. 47) of the polysaccharides. The presence of the same methylated sugars were detected from all three polysaccharides confirming the presence of arabinose 3-sulphate (see Fig. 7).

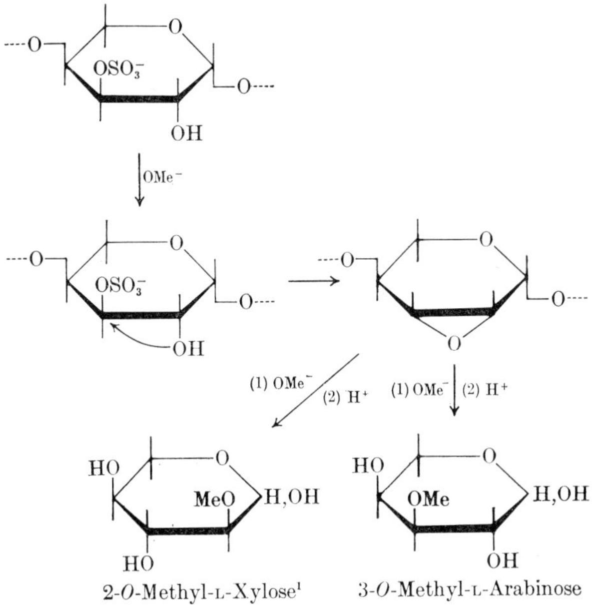

2-O-Methyl-L-Xylose[1]   3-O-Methyl-L-Arabinose

FIG. 7.

[1] Characterized as crystalline material from C. rupestris.

C. CAULERPA FILIFORMIS

The water-soluble polysaccharides of *C. filiformis* also comprise complex sulphated material together with a small proportion of starch (see p. 79) (Mackie and Percival, 1961). Three samples of this weed collected from the same site in South Africa were examined. Whereas two samples harvested in February, 1957 and 1958, were devoid of arabinose-containing polysaccharide, this sugar was present in a sample collected in November, 1958. It should be pointed out, however, that the more drastic conditions, dilute acid at 70° used in the extraction of the February samples as compared with cold water for the November sample, may explain this difference, although acid treatment of the polysaccharide isolated from the November sample failed to remove the arabinose. Furthermore, examination of the water-soluble polysaccharides of *Caulerpa racemosa* and of *C. sertularoides* collected in October from dead coral in strong surf on the Gata Islands showed the complete absence of arabinose residues although the proportions of the other sugars were similar to those of the *Caulerpa filiformis* extracts (Table 3) and roughly in the same proportions. Although the absence of arabinose may be a seasonal or environmental variation, until arabinose-containing polysaccharides have been found in other samples of *Caulerpa* spp. they cannot be regarded as characteristic of this genus of green seaweed.

After removal of the glucan with salivary α-amylase, the respective materials isolated from *C. filiformis* had the following properties and composition (Mackie and Percival, 1961). Polysaccharide (F) from weed collected in February had $[\alpha]_D$ +10·7° and contained 16·4% sulphate, 14·0% ash, 19·5% protein; and (N) from weed collected in November had $[\alpha]_D$ +12·8°, 17·6% sulphate, 15·2% ash and about 19% protein. The proportions of the sugars in acidic hydrolysates are given in Table 3.

Table 3.

| | Approximate molar proportions of | | | |
	Galactose	Mannose	Xylose	Arabinose
Polysaccharide (F)	5	2	1	0
Polysaccharide (N)	5	2	2	1
Small quantities of glucose and rhamnose were also detected.				

A number of fractionation techniques failed to give separation into homopolysaccharides, but partial fractionation was obtained by dropwise addition of a saturated solution of barium hydroxide to a solution of the free acid polysaccharide (Meier, 1958), and removal of precipitates from time to time.

The composition of the different fractions compared with that of the original material are given in Table 4.

Table 4   Fractionation with Barium Hydroxide

	Wt. mg	Sulphate	$[\alpha]_D$	Galactose	Arabinose	Xylose	Mannose
						Sugars (%)	
Polysac. (N)	500	17·6	+12·8°	49·4	9·2	18·9	22·5
Fraction 1	128	17·4	12·6	46·1	6·9	21·5	25·5
Fraction 2	93	18·1	12·7	61·3	12·0	13·6	13·1
Fraction 3	222	18·3	13·2	62·2	12·1	13·2	12·5

Failure to separate a homopolysaccharide from this material is by no means proof of its absence and the partial fractionation indicates at least that there is more than a single polysaccharide present in the extract. In view of the fact (Mackie and Percival, 1959) that the major structural polysaccharide synthesized by this green seaweed is a β-1,3-linked xylan (see p. 91), it is possible that the xylose containing material in the present extracts is the low molecular weight water-soluble part of the xylan which is extracted from the residual weed with alkali.

Apart from the separation of a small quantity of a galactose sulphate from a partial acid hydrolysate, no evidence for the site of ester sulphate or of the linkages between the sugars in this extract has yet been advanced.

D. CODIUM FRAGILE

Three samples of *C. fragile* have been investigated: two collected from South Africa in February and October, respectively, and a third from Biarritz, France, in September. Qualitative examination showed that all three samples synthesized the same polysaccharides in roughly the same proportions (Love and Percival, 1964). Cold and hot water extraction of this green seaweed gave essentially similar polysaccharide material, although the content of glucose-containing polysaccharide was higher in the latter. After separation of a starch-type polysaccharide as the starch–iodine complex (see p. 79) from the hot water extract, the residual solution was combined with the cold water extract which it closely resembled. After dialysis of the combined solutions, the polysaccharide was isolated as a white powder, from the solution remaining in the dialysis sac, by freeze-drying. It had $[\alpha]_D$ +46° and comprised galactose and arabinose together with lesser amounts of mannose, xylose and glucose and trace amounts of rhamnose together with 17% sulphate (as SO_3^{2-}) and 6·2% uronic acid (by decarboxylation) and 25% of contaminating protein. Although chromatographic evidence for the presence of uronic acid was obtained, all attempts to isolate and characterize it were unsuccessful. It must be remembered that the pentoses no doubt contributed some of the carbon dioxide in the decarboxylation (see p. 30) and the amount of uronic acid actually present is probably very small. The small proportion of glucose was derived from residual starch since this could be completely eliminated by salivary α-amylase. The cell wall of this alga comprises a β-1,4-mannan (see p. 94) and the small amount of mannose containing polysaccharide in this extract is probably low molecular weight mannan derived from the cell walls. The fact that this is completely removed by periodate oxidation supports this conclusion.

Evidence of the heterogeneity of the material was obtained by elution from a column of DEAE-cellulose with increasing concentration of potassium chloride. The major fraction (A) was refractionated on a fresh column and fractions with increasing sulphate content and less mannose and xylose were separated as the concentration of potassium chloride increased and the final fraction contained only arabinose, galactose and sulphate (Table 5).

Table 5   Fractionation of water-soluble polysaccharide material
from *Codium fragile.*

	$[\alpha]_D$	Sulphate	Galactose	Mannose	Arabinose	Xylose
Fraction (A)	$+37°$	12·7%	*****	**	***	**
Fraction 1 (7%)	—	—	—	***	—	***
Fraction 2 (18%)		11%	***	*	—	*
Fraction 3 (47%)	$+45°$	19%	*****	—	***	*
Fraction 4 (27%)	$+55°$	29%	****	—	****	—

**** = major component; * = trace.
About a third of the material was irreversibly bound on the column.

Partial hydrolysis of the original extract, $[\alpha]_D$ $+46°$ led to the separation and characterization of 3-*O*-β-D-galactopyranosyl-D-galactose [Fig. 8 (**23**)] and 3-*O*-β-L-arabinopyranosyl-L-arabinose [Fig. 8 (**24**)]. This indicates that a proportion of the arabinose and galactose occur as adjacent 1,3-linked units in the polysaccharide. Supporting evidence for this was given by the presence of these two sugars in the hydrolysate of the periodate oxidized polysaccharide (see p. 35).

After fractionation of the partial acid hydrolysate on anion exchange resin, two galactose monosulphates were also isolated. These were characterized as galactose 4- [Fig. 8 (**25**)] and 6- [Fig. 8 (**26**)] monosulphates, respectively. The presence of the latter was confirmed by the formation of 2% of 3,6-anhydrogalactose (see p. 47) and removal of about 4% of the sulphate on

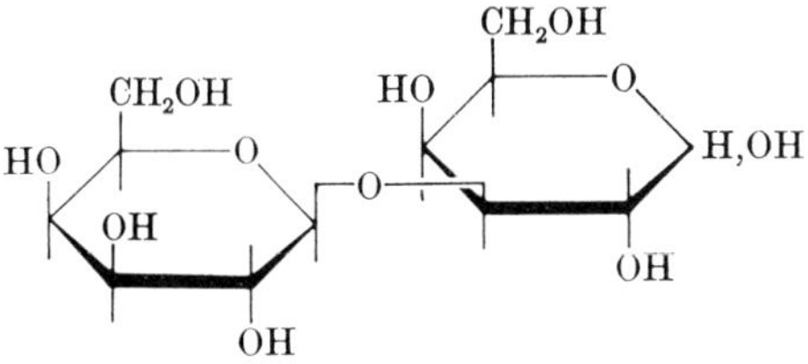

CH$_2$OH

CH$_2$OH   HO

HO

H,OH

OH

OH

OH

(23)

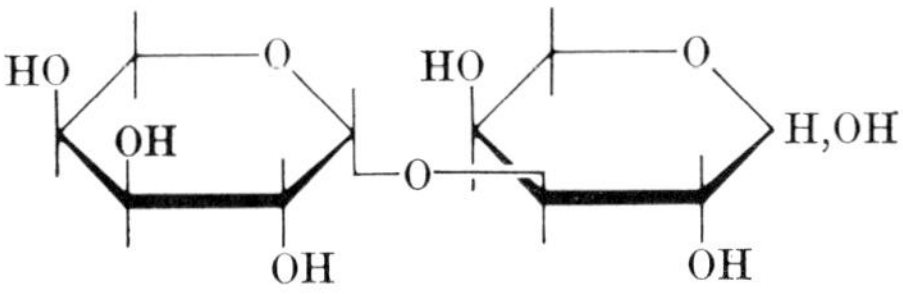

HO   HO

OH

H,OH

OH

OH

(24)

CH$_2$OH   CH$_2$OSO$_3^-$

$^-$O$_3$SO   HO

OH   H,OH   OH   H,OH

OH   OH

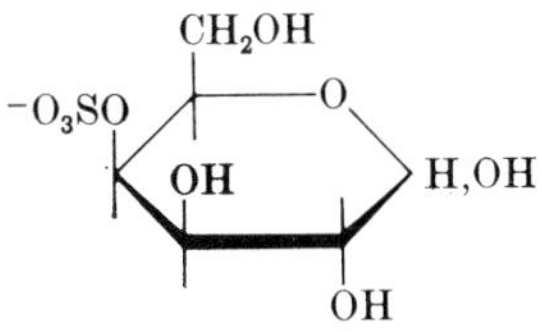

(25) 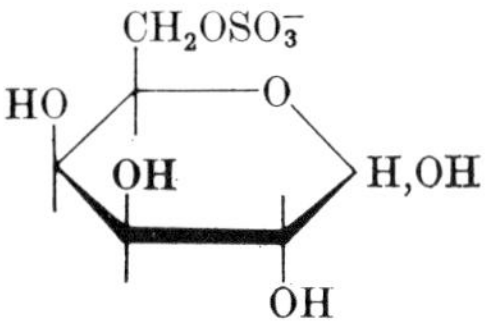   (26)

Fig. 8.

Table 6   Water-soluble sulphated polysaccharides from:

	Cladophora rupestris	*Chaetomorpha capillaris*	*Caulerpa*[1] *filiformis*	*Codium fragile*
$[\alpha]_D$	+69°	+66°	+13°	+46°
Approx. mol. ratio Gal:Ara:Xyl	3:3·5:1	1·5:3·7:1	2·5:0·5:1	2:2:1
Other sugars present in small amounts	rhamnose glucose	rhamnose glucose	mannose[2] glucose rhamnose	rhamnose
Approx. protein %	8	19	19	25
Sulphate as SO$_4^{2-}$ %	15	15	17	20
Site of sulphate	Gal-6S Ara-3S	[Gal-6S] [Ara-3S]	Gal	Gal-4S Gal-6S
Disaccharides from partial hydro-lysates	Gal(1 → 3)Gal Gal(1 → 6)Gal Ara(1 → 4/5)Ara	[Gal(1 → 3)Gal] [Gal(1 → 6)Gal] [Ara(1 → 4/5)Ara]	—	Gal(1 → 3)Gal Ara(1 → 3)Ara

[1] Arabinose appeared to be absent from most samples.
[2] Mannose was present in approximately 1 molar proportion.
The presence of square brackets indicates chromatographic identification only.

treatment of the polysaccharide with alkali. Since the proportion of the galactose to arabinose in the alkali-treated material was the same as in the original extract, it is probable that the alkali also removed sulphate from the arabinose with epoxide formation and subsequent conversion into xylose (see p. 46). Infrared bands given by the polysaccharide at 850 cm^{-1} and 820 cm^{-1} confirmed the presence of the two galactose sulphates (see p. 48).

The properties and constituents of these sulphated extracts from the Chlorophyceae are compared in Table 6.

In some respects these materials appear to be very similar. They all have positive rotations, contain similar sugars and proportion of ester sulphate and in each of them a proportion of the latter is linked to galactose. They appear to have highly branched molecules with a proportion of 1,3- and 1,6-linkages.

REFERENCES

Bernardi, G., and Springer, G. F. (1962). *J. biol. Chem.* **237**, 75.

Black, W. A. P. (1954). *J. Sci. Fd. Agric.* **5**, 445.

Black, W. A. P., Cornhill, W. J., Dewar, E. T., Percival, E. G. V., and Ross, A. G. (1950). *J. Soc. chem. Ind., Lond.* **69**, 317.

Black, W. A. P., Dewar, E. T., and Woodward, F. N. (1952). *J. Sci. Fd. Agric.* **3**, 122.

Black, W. A. P., Cornhill, W. J., Dewar, E. T., and Woodward, F. N. (1953). *J. Sci. Fd. Agric.* **4**, 85.

Cameron, M. C., Ross, A. G., and Percival, E. G. V. (1948). *J. Soc. chem. Ind. Lond.* **67**, 161.

Conchie, J., and Percival, E. G. V. (1950). *J. chem. Soc.* p. 827.

Côté, R. H. (1959). *J. chem. Soc.* p. 2248.

Dillon, T., Kristensen, K., and O'hEocha, C. (1953). *Proc. R. Irish Acad.* **55B**, 189.

Duncan, W. A. M., Manners, D. J., and Ross, A. G. (1956). *Biochem. J.* **63**, 44.

Fisher, I. S., and Percival, Elizabeth (1957). *J. chem. Soc.* p. 2666.

Fromageot, C., and Heitz, P. (1938). *Mikrochem. Acta* **3**, 52.

Galli, D. R., and Giese, A. C. (1959). *J. exp. Zool.* **140**, 415.

Hirst, Sir Edmund, Mackie, W., and Percival, Elizabeth (1965). *J. chem. Soc.* p. 2958.

Kylin, H. (1913). *Hoppe-Seyler's Z. physiol. Chem.* **83**, 171.

Kylin, H. (1915). *Hoppe-Seyler's Z. physiol. Chem.* **94**, 357.

Larsen, B., and Haug, A. (1963). *Acta chem. scand.* **17**, 1646.

Larsen, B., Haug, A., and Painter, T. L. (1966). *Acta chem. scand.* **20**, 219.

Lloyd, K. O. (1960). Ph.D. Thesis, University of Wales.

Love, J., and Percival, Elizabeth (1964). *J. chem. Soc.* p. 3338.

Love, J., Mackie, W., McKinnell, J. P., and Percival, Elizabeth (1963). *J. chem. Soc.* p. 4177.

Lunde, G., Heen, E., and Oy, E. (1937). *Hoppe-Seyler's Z. physiol. Chem.* **247**, 189.

McCulley, M. E. (1965). *Can J. Bot.* **43**, 1001.

Mackie, I. M., and Percival, Elizabeth (1959). *J. chem. Soc.* p. 1151.

Mackie, I. M., and Percival, Elizabeth (1961). *J. chem. Soc.* p. 3010.

McKinnell, J. P., and Percival, Elizabeth (1962). *J. chem. Soc.* p. 2082.

McNeely, W. H. (1959). *In* "Industrial Gums" (R. L. Whistler and J. N. BeMiller, eds), Academic Press, New York and London.

Meier, H. (1958). *Acta chem. scand.* **12**, 144.

O'Donnell, J. J., and Percival, Elizabeth (1959). *J. chem. Soc.* p. 1739.

O'Neill, A. N. (1954). *J. Am. chem. Soc.* **76**, 5074.

Nelson, W. L., and Cretcher, L. H. (1931). *J. biol. Chem.* **94**, 147.

Percival, E. G. V. (1949). *Q. Rev. chem. Soc.* **3**, 369.

Percival, E. G. V., and Ross, A. G. (1950). *J. chem. Soc.* p. 717.

Percival, Elizabeth (1962). *In* "Methods of Carbohydrate Chemistry" (R. L. Whistler and M. L. Wolfrom, eds), Vol. 1, p. 197, Academic Press, New York and London.

Percival, Elizabeth (1967). *Chemy. Ind.* 511.

Percival, Elizabeth, and Wold, J. K. (1963). *J. chem. Soc.* p. 5459.

Quillet, M. (1961). *Colloques int. Cent. natn. Rech. scient.* **103**, 10.

Schuler, W., and Springer, G. F. (1957). *Naturwissenschaften* **44**, 26.

Schweiger, R. G. (1962a). *J. org. Chem.* **27**, 4267.

Schweiger, R. G. (1962b). *J. org. Chem.* **27**, 4270.

Springer, G. F., Wurzel, H. A., McNeal, G. M. Jr., Ansell, N. J., and Doughty, M. F. (1957). *Proc. Soc. exp. Biol. Med.* **94**, 404.

Yaphe, W. and Morgan, K. (1959). *Nature, Lond.* **183**, 761.

CHAPTER 8

Polysaccharides Containing Uronic Acid and Ester Sulphate

Uronic acid-containing sulphated polysaccharides have been reported from all the major classes of seaweeds although the extent of chemical investigations on these polymers is relatively small.

I. POLYURONIDES OF THE PHAEOPHYCEAE

In a recent investigation of *Ascophyllum nodosum*, the Norwegian School (Haug and Larsen, 1963; Larsen *et al.*, 1966) found that after removal of acid soluble carbohydrates they could extract with dilute alkali from this weed a mixture of alginic acid and glucuronoxylofucans. The alginic acid could be precipitated as the calcium salt and the other polysaccharides fractionally precipitated with alcohol. The major fraction (6% of the dry weight of weed), termed by the authors "ascophyllan", contained approximately 25% fucose, 26% xylose, 19% sodium uronate, 13% $NaSO_3'$ and 12% protein. Two further fractions F_1 and F_2 (each about 1·5% of the dry weight) were also isolated. These differed from ascophyllan in the proportions of the constituents and F_2 more closely resembles fucoidan (see p. 157) in its higher (38·6%) fucose content. All the fractions were brown in colour due, it is considered, to contamination with condensation products of phenols with the polypeptide, since previous treatment of the alga with formaldehyde gave colourless fractions (Haug, 1964).

Ascophyllan and F_2 were shown to be electrophoretically homogeneous materials, whereas F_1 contained a small proportion of a second material. All three fractions gave the same chromatographic pattern on autohydrolysis in the free acid form, although there were considerable differences in the intensities of the respective spots.

The monosaccharides in ascophyllan were separated from a hydrolysate and characterized as L-fucose, D-xylose, D-glucuronic acid, and eighteen common amino acids derived from the polypeptide portion were identified by two-way chromatography.

All attempts to separate the polypeptide from the carbohydrate were unsuccessful and this, together with the electrophoretic homogeneity of asco-phyllan, suggests a chemical linkage between the peptide and the carbo-hydrate which could be cleaved by heating in the acid form (pH 2·05) at 80° for 20 hr.

This method of hydrolysis, and also hydrolysis with 0·5N-oxalic acid at 100°, besides cleaving the polypeptide link also hydrolysed the polysac-charide into dialysable oligosaccharides containing xylose and fucose and a non-dialysable material which remained in solution when the polypeptide precipitated. The non-dialysable degraded polysaccharide comprised almost all the uronic acid present in the original ascophyllan and was almost devoid of fucose, xylose and half-ester sulphate. From these results the authors de-duce that the glucuronic acid comprises the backbone of the macromolecule to which are attached glycosidically, relatively long, side-chains of sulphated fucose and xylose. On the available evidence the polypeptide appears to be linked to the fucose by more than a single type of linkage.

Further support for the suggestion that the components are covalently linked together in a single complex macromolecule was provided by the fact that "alkali-treated" ascophyllan liberates dialysable hexuronic acid and peptides when subjected to the same hydrolytic conditions as the untreated material. It was established that the alkali cleaved internal linkages in both the polyuronide and polypeptide chains but the derived fragments of uronides remained attached to the fucose until hydrolysed by the acid.

Polysaccharide material comprising the same sugars but of different struc-ture was extracted from weed residues of *A. nodosum* remaining after lami-naran, fucoidan and alginic acid had been extracted in the usual way. The residual weed was first treated with formaldehyde and then extracted at 80° with ammonium oxalate and oxalic acid at pH 2·8. The extract contained about 10% of alginic acid and this was separated as the insoluble calcium salt. The remaining polysaccharide material, isolated as a cream solid (about 22% of the dry weight of *Ascophyllum nodosum*) by freeze-drying, had $[\alpha]_D$ −225°. It comprised fucose (5 parts), xylose (1 part) and glucuronic acid (1 part); and 20% sulphate and 3·8% protein (Percival, 1967). Like the ascophyllan described above, it defied fractionation into homopolysaccharides although some variation in the proportion of the constituents was achieved by frac-tionation with quaternary ammonium salts, sodium chloride and ethanol. Besides differing from ascophyllan in the proportions of its constituents (see Table 1), it appears to have a different macromolecule. After autohydrolysis of the free acid form, the degraded polysaccharide, $[\alpha]_D$ −87°, recovered in about 17% yield contained all the initial sugars, fucose (3·5 parts), xylose (1 part) and glucuronic acid (2·5 parts). Unlike ascophyllan, the glucuronic acid does not appear to form the backbone of the molecule and a xylosylfucose and a fucosylxylose together with a number of higher

Table 1   Approximate percentage composition.

	Fucose	Xylose	Sodium uronate	NaHSO₃	Protein
Ascophyllan	25	26	19	13	12
Glucuronoxylofucan from weed residues	49	10	12	21	4
Degraded glucuronoxylofucan	42	12	29	12	5

oligouronides were present in the hydrolysate. Similar treatment of the weed residue of the stipe of *Laminaria hyperborea* gave a lower yield of the same type of polysaccharide material.

These studies indicate that the type and variety of algal fucose-containing polysaccharides is much wider than originally believed. The name fucoidan is generally accepted to mean polysaccharides consisting almost entirely of fucose and ester sulphate and until more genera of Phaeophyceae have been examined for their fucose-containing polysaccharides, the present authors consider it undesirable to extend the name fucoidan to include the complex polymers discussed in this chapter which are more systematically named glucuronoxylofucan sulphates.

II. POLYURONIDES OF THE RHODOPHYCEAE

Glucuronic acid has been reported as a constituent of agaropectin (see p. 133), but no structural studies in which glucuronic acid plays a part have been reported for this polysaccharide.

In the mucilage extracted by dilute acid from *D. edulis*, 9·5 to 11% of uronic acid is reported and an analysis showed one uronic acid to nine galactose units and two ester sulphate groups. The latter were stable to alkali (see p. 46). Barry degradation (see p. 41) indicated the presence of 1,3- and 1,6-linked galactose units (Barry and Dillon, 1945). However, while methylation of a degraded acetate confirmed the presence of a high proportion of 1,3-linked galactose units, no evidence for 1,6-linkage nor any indication of the role of the uronic acid in the macromolecule was obtained.

III. POLYURONIDES OF THE CHLOROPHYCEAE

A. INTRODUCTION

Two members of the family Ulvaceae, *Ulva lactuca*, and *Enteromorpha compressa*, and a member of the family Acrosiphonaceae, *Acrosiphonia arcta* (*A. centralis, Spongomorpha arcta*), of the green seaweeds have been shown to synthesize similar water-soluble sulphated polyuronides. *U. lactuca* harvested from the lower bay of Fundy, Nova Scotia, the east coast of Scotland, and from Kames Bay, Millport, Isle of Cumbrae, Scotland, all gave the same water-soluble sulphated mucilage. *Enteromorpha compressa* and *Acrosiphonia arcta* were collected from the shores near Millport. The different species of these latter weeds are difficult to distinguish and considerable care is necessary in the collection of a single species. *Acrosiphonia* is morphologically closely related to the Cladophoras, but appears, at least with regard to its carbohydrates, to have a different metabolic system (cf. Table 2 and p. 165). All three mucilages extracted from the respective weeds with hot water were contaminated with starch when first extracted but this can be separated as the starch–iodine complex, and the residual materials all contain the same sugars and have many similar properties (see Table 2) and give almost identical infrared spectra.

Table 2   Water-soluble polysaccharides from:

	Acrosiphonia arcta[1]	*Enteromorpha compressa*[2]	*Ulva lactuca*[3]
$[\alpha]_D$	$(-31°)$*	$(-47°) -87°$	$(-47°) -74°$
Equiv. of free acid polysaccharide	459	310	355
Constituent sugars:			
L-Rhamnose	***	*****	*****
D-Xylose	***	***	***
D-Glucose	*	*	*
D-Glucuronic acid %	19	18	20
Sulphate %	7·8	16	17·5
Mole 10$'_4$ reduced/anhydro unit	1·0	0·38	0·30

* Figures in brackets are rotations before separation of contaminating starch.
[1] O'Donnell and Percival (1959).
[2] McKinnell and Percival (1962b).
[3] Brading *et al.* (1954); Percival and Wold (1963).

It is unlikely that the glucose is derived from residual starch since the polysaccharides no longer gave a colour with iodine and were unattacked by salivary α-amylase.

The most striking differences between the mucilages appears to be the lower rhamnose and sulphate content of *A. arcta*, but this may be a seasonal or environmental effect. At the same time, it must be remembered that the presence of uronosyl linkages makes complete hydrolysis, without degradation, impossible and for this reason no accurate determination of the proportions of neutral sugars in such polysaccharides can be made. After 6 and 16 hr hydrolysis of *U. lactuca* mucilage, the relative molar proportions were found to be glucose:xylose:rhamnose 1:3·4:4·8 and 1:3·5:6·9, respectively, together with trace quantities of galactose and mannose, and the hydrolysates of *A. arcta* usually contained about 45% of oligouronic acids.

The extracts gave negative tests for the presence of ester lactone, keto-sugar, 3,6-anhydrogalactose and amino sugar (Cessi and Piliego, 1960). All three polysaccharides contain a proportion of protein, that of an alkali extract of *U. lactuca* being about 25% (Brading *et al.*, 1954), but in the aqueous extracts this was considerably lower (about 2·5% for *U. lactuca*, 5·5% for *E. compressa* and 4% for *A. arcta*).

Attempts to separate the extracts into homopolysaccharides by fractional precipitation with quaternary ammonium salts (O'Donnell and Percival, 1959; McKinnell and Percival, 1962b), preferential precipitation with copper salts or barium hydroxide were unsuccessful. Elution of the *Ulva* mucilage from a column of DEAE-cellulose with increasing concentration of potassium chloride gave three fractions which contained approximately the same relative proportion of monosaccharides and gave identical infrared spectra, but they had limiting viscosity numbers of 1·5, 1·10 and 0·95 (Percival and Wold, 1963), respectively. The sedimentation patterns of the original polysaccharide and fraction 3 examined in the ultracentrifuge indicated heterogeneity in the former, while fraction 3 gave a single sharp peak. From these facts it may be concluded that these mucilages are polydisperse heteropolysaccharides.

B. STRUCTURE

1. *Methylation Studies*

Although the high acidity of these mucilages made complete methylation difficult, evidence for the presence of endgroup xylose, 1,4-linked xylose and rhamnose and of triply linked rhamnose was obtained from methylation studies on *U. lactuca* (Brading *et al.*, 1954) and on *A. arcta* (O'Donnell and Percival, 1959). To overcome the methylation difficulties, *U. lactuca* polysaccharide was converted into nearly neutral material by desulphation with methanolic hydrogen chloride and reduction of the carboxyl groups with borohydride (Haq and Percival, 1966). The recovered, partly degraded polysaccharide had $[\alpha]_D$ $-75°$ and contained SO_3^{2-} 1·3% and uronic anhydride 5·4%. After exhaustive methylation the product had a methoxyl content of 33·4%. From the characterization of the various methylated sugars, in addition to the linkages already established, the presence of end group glucose, 1,3-linked glucose and triply linked xylose was proved. Tentative evidence for 1,3-linked xylose and 1,4-linked glucose was also obtained. It must be borne in mind that a large proportion of the glucose residues must be derived from glucuronic acid in the native polysaccharide.

2. *Partial Hydrolysis*

Partial acid hydrolysates from all three mucilages contained a high proportion of an aldobiouronic acid which was separated and characterized as 4-*O*-β-D-glucopyruronosyl-L-rhamnose (**1**) (O'Donnell and Percival, 1959; McKinnell and Percival, 1962a).

(1)

In addition, from *U. lactuca* small quantities of two glucuronosylxyloses, thought to be glucuronosyl($1 \rightarrow 3$)xylopyranose and glucuronosyl($1 \rightarrow 4$)xylopyranose, were separated, and tentative evidence for the presence of the tetrasaccharide:

GlcA($1 \rightarrow 4$)Rha($1 \rightarrow 3/4$)GlcA($1 \rightarrow 3$)Xyl

GlcA = D-glucuronic acid, Rha = L-rhamnose, Xyl = D-xylose

was also obtained (Haq and Percival, 1966a).

Furthermore, hydrolysis of the desulphated carboxyl-reduced polysaccharide from *U. lactuca* gave a mixture of oligosaccharides from which Rha($1 \rightarrow 4$)Xyl($1 \rightarrow 3$)Glc was separated and characterized. These results confirm the presence of the linkages indicated by methylation studies and provide the first evidence that the three sugars are present in a single polysaccharide and, for some of the structural units, present in the macromolecule.

3. *Application of Smith Degradation* (see p. 39)

The mucilage from *U. lactuca* was oxidized with periodate and the resulting polyaldehyde reduced. The polyalcohol was recovered in 73% yield and on mild hydrolysis gave 7·2% of degraded polyalcohol and 50% of fragments (Haq and Percival, 1966b). Hydrolysis of the former gave the characteristic aldobiouronic acid, rhamnose, xylose and a number of spots on a paper chromatogram which were not identified. Re-oxidation of the polyalcohol reduced 0·12 moles of periodate per C_6 anhydro unit and the hydrolysate of the recovered polyalcohol gave the same chromatographic pattern as that of the first degraded polyalcohol.

The fragments consisted of glycerol (**10**), glycollic aldehyde (**8**) and five non-reducing carbohydrates which were separated and characterized as erythritol (**7**) (8·7%), 2-*O*-D-glucosyl-erythritol (**9**) (10·5%), 2-*O*-β-D-xylosyl-glycerol (**11**) (10·5%) and 2-*O*-D-xylosylerythritol (**12**) (8·7%), and an acidic oligosaccharide (61%). Erythritol (**7**) and glycollic aldehyde (**8**) are derived from 1,4-linked glucose units (**2**) and glycerol (**10**) from 1,4-linked xylose (**4**). The 2-*O*-β-D-glucosylerythritol (**9**), 2-*O*-β-D-xylosyl-glycerol (**11**) and -erythritol (**12**) are derived as shown in Fig. 1, although the units are not necessarily linked in this sequence in the native polysaccharide. An alternative structure (**13**) which yields the same products is also given in Fig. 1.

7

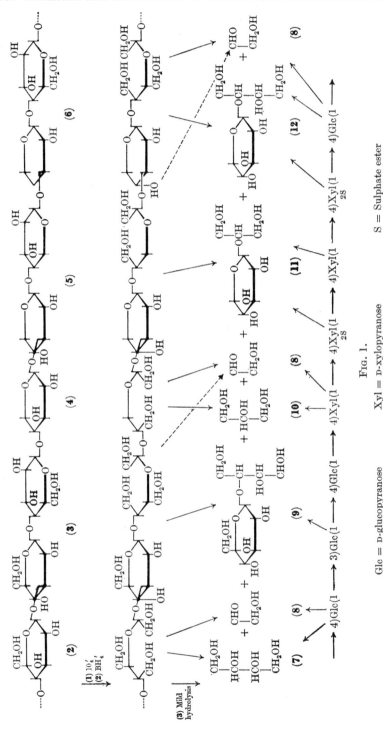

FIG. 1.

Glc = D-glucopyranose   Xyl = D-xylopyranose   S = Sulphate ester

β-linkage is depicted because of negative rotation of the polysaccharide.
For site of ester sulphate see later section.

The acidic oligosaccharide, $[\alpha]_D$ $-13°$ (about two thirds of the fragments) had a chromatographic mobility ($R_{\text{Galacturonic acid}}$ 0·17) of a heptasaccharide and an ionophoretic mobility ($M_{\text{Galacturonic acid}}$ 1·3) of a highly-charged molecule. It contained about 25% ester sulphate (as SO_3^{2-}) and had an equivalent of 245, and on hydrolysis gave erythronic acid, glucose, xylose, rhamnose, 4-O-glucuronosylrhamnose and glycerol. The presence of erythronic acid, which could only have arisen from 1,4-linked glucuronic acid, was unexpected. The fact that it was not cleaved on mild hydrolysis is probably due to the stabilizing effect of the carboxyl group on the acetal linkage. Methylation of this fragment and gas-liquid chromatography of the methanolysate indicated the presence of end group glucuronic acid, 1,4-linked rhamnose, and 1,3-linked glucose and xylose with sulphate groups on rhamnose, xylose and glucose. A structure [Fig. 2 (14)] which embodies all these facts is given below:

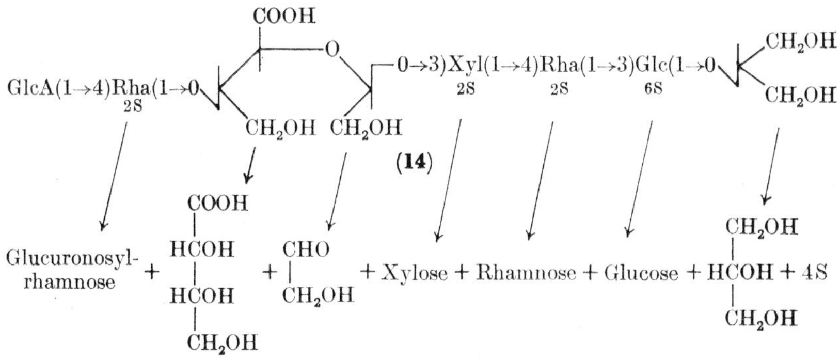

GlcA = D-glucuronic acid   Rha = L-rhamnopyranose   S = Sulphate ester
Glc = D-glucopyranose   Xyl = D-xylopyranose

For evidence for the site of ester sulphate see next section.

FIG. 2.

The tetrasaccharide tentatively identified in the partial acid hydrolysate could be derived from the first four units in this fragment. In view of the high yield of this oligosaccharide the unit

$$\text{GlcA}(1 \to 4)\underset{2S}{\text{Rha}}(1 \to 4)\text{GlcA}(1 \to 3)\text{Xyl}(1 \to 4)\underset{2S}{\text{Rha}}(1 \to 3)\text{Glc}(1 \to 4)\text{Xyl}$$

or similar structure in which the order of the units may be different must occur as a repeating structure in the macromolecule.

4. The Site of Ester Sulphate

It was found that the amount of periodate reduced by the polysaccharides from *E. compressa* and *U. lactuca* was virtually doubled after desulphation of the respective polysaccharides. From this it was deduced that the removal of sulphate ester groups from the polysaccharide had led to the formation of additional adjacent hydroxyl groups. Examination of hydrolysates of the various oxopolysaccharides revealed considerably less uncleaved rhamnose units in the material derived from the desulphated polysaccharides, indicating that sulphate had been removed from rhamnose. This was confirmed by allowing the oxidation with periodate to proceed at 2° in buffered solution; again the desulphated material reduced twice as much oxidant. Under these conditions, vicinal *cis*-hydroxyl groups in sugars are oxidized preferentially to *trans*-glycol groups. Since rhamnose is the only sugar in this polysaccharide which has *cis*-hydroxyl groups (at C-2 and C-3), it follows that the increased periodate reduction of the desulphated material is probably due to removal of sulphate at C-2 and/or C-3 of rhamnose. Such sulphate would be stable to alkali (see p. 46) which is the case for the majority of the ester sulphate in this polysaccharide. Infrared examination of the polysaccharide revealed a broad peak at 1240 cm^{-1}, characteristic of the S$=$O stretching vibration, and a second peak at 850 cm^{-1}, characteristic of axial sulphate in galactose units. A large proportion of the ester sulphate was, therefore, tentatively assigned to C-2 of the rhamnose residues on the assumption that the infrared results for galactose can be applied to L-rhamnose which, it is assumed, is present in the polysaccharide in its more stable 1C conformation in which only C-2 is axial.

A 71% recovery of the *Ulva* polysaccharide in which the sulphate had been reduced from 14·1 to 12·5%, was achieved after treatment with N-sodium hydroxide at 80° for 14 hr in the presence of borohydride. This latter reagent converts the reducing end-groups into sugar alcohols and so prevents alkaline degradation of the polysaccharide. A hydrolysate of the recovered material contained, in addition to acidic fragments, glucose, xylose, rhamnose, and arabinose in the molar proportions of 1:1·6:2·0:0·1 and trace amounts of mannose, galactose and lyxose. Neither arabinose nor lyxose were present in the hydrolysate of the initial polysaccharide. The sugars, apart from galactose and lyxose, were separated and characterized as L-rhamnose, D-xylose, D-mannose and D-arabinose. The separated syrupy lyxose had a negative rotation and was therefore probably D-lyxose. The arabinose units formed as a result of alkali treatment of the polysaccharide can only have arisen from xylose, since none of the other sugars could possibly have been converted into this sugar under the conditions of the experiment. Its formation can be explained if the polysaccharide contains 1,4-linked xylose residues which are sulphated on C-2 (**15**) or C-3 (**16**) (see p. 46). Such residues under the action of alkali yield epoxide derivatives which under the further action of alkali and acid hydrolysis of the polysaccharide give D-arabinose (**17**) and/or D-xylose (**18**) units. The removal of about a tenth of the sulphate by alkali corresponds to monosulphation of about 15% of the xylose units.

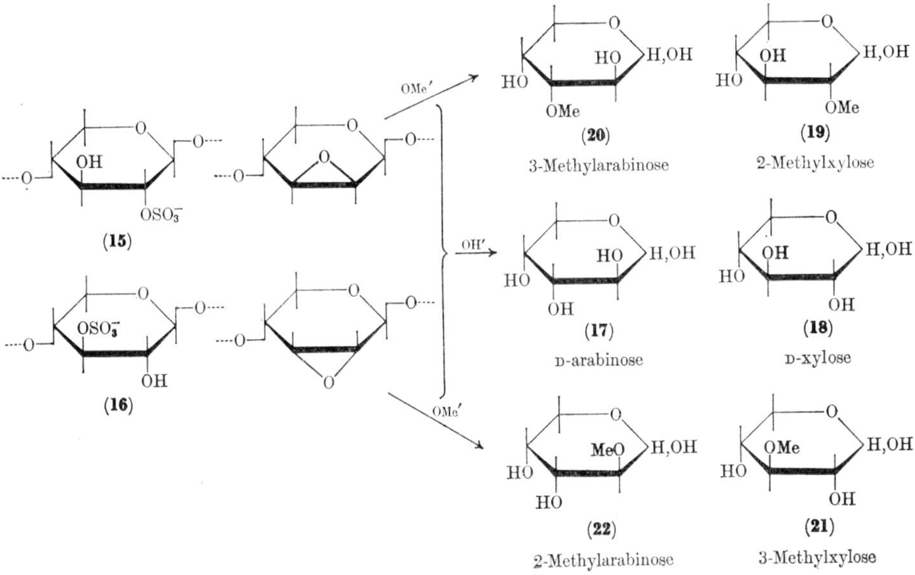

Fig. 3.

The formation of D-arabinose does not distinguish between 2- and 3-sulphated xylose units. In contrast, cleavage of epoxide rings with sodium methoxide yields monomethyl sugars, the methoxyl group entering the sugar unit at the site of attack. Consequently, two methyl sugars are possible from each epoxide ring as is shown in the scheme Fig. 3. Xylose 2-sulphate (**15**) should yield 2-*O*-methylxylose (**19**) and 3-*O*-methylarabinose (**20**), whereas xylose 3-sulphate (**16**) gives 3-*O*-methylxylose (**21**) and 2-*O*-methylarabinose (**22**). Because of steric factors, only the one methyl sugar is frequently formed in any quantity. In the present experiments, 2-*O*-methylxylose (**19**) was the sole methylated sugar to result. The trace of D-lyxose can be explained by epoxide ring migration (see p. 47) and could have arisen from end-group xylose sulphated at C-2. The only evidence for the glucose 6-sulphate depicted in Fig. 2 is the resistence to acid hydrolysis of a small portion of the sulphate (see Rees, 1963).

5. Conclusions

Although the investigations on the mucilage from *U. lactuca* have advanced much further than those on the mucilages from *E. compressa* and *A. arcta*, the essential similarity of the three materials has been established. All three are highly branched, polydisperse heteropolysaccharides which contain D-glucuronic acid, L-rhamnose, D-xylose and D-glucose and carry ester sulphate on rhamnose and probably a small proportion on xylose, and they all contain the structural unit 4-O-β-D-glucuronosyl-L-rhamnose (**1**).

It has also been established for the *Ulva* mucilage that all the sugars are indeed present in a single molecule, that adjacent xylose and adjacent glucose units are a feature of the structure and that the inner core of the macromolecule appears to comprise the same sugars as the periphery since the degraded polysaccharides, recovered after partial hydrolysis in about 5% yield from the nearly neutral polymer and in 6 to 10% yield from the original polysaccharide, contain all the sugars found in the initial material.

As knowledge increases, it is probable that differences in the fine structure of the mucilages from the different genera will emerge.

IV. POLYURONIDE OF THE BACILLARIACEAE

After exhaustive extraction of a sample of the marine diatom, *Phaeodactylum tricornutum* (see p. 70) grown under bacteria-free conditions, with hot water, mild chlorite and cold alkali, the residual material which comprised about 5% of the dry weight of the organism almost completely dissolved in hot 4% alkali under nitrogen (Ford and Percival, 1965). Addition of ethanol, after neutralization of the hot alkali extract, dialysis and concentration, precipitated a glucuronomannan as a white powder, $[\alpha]_D$ +34°. In contrast to the previous mannans (see p. 93), the present material was completely soluble in water and contained in addition to mannose, 27% glucuronic acid and 7·5% half ester sulphate. Acidic hydrolysis gave mannose, glucuronic acid and three oligouronic acids, two of which were characterized as O-D-glucopyranosyluronic acid $(1 \rightarrow 3)$D-mannopyranose (**23**) and O-D-glucopyranosyluronic acid $(1 \rightarrow 3)$D-mannopyranosyl$(1 \rightarrow 2)$D-mannopyranose (**24**), respectively. The third acid contained some six units which on hydrolysis gave mannose, glucuronic acid, glucose and the above di- and tri-oligouronic acids.

(**23**)   (**24**)

(No evidence for α- or β-linkage was advanced)

The application of Smith degradation (see p. 39) to this polysaccharide led to the recovery of about 25% of a degraded polymer containing only mannose units and about 10% sulphate. Glycerol (29), glyceric acid (27), several slower spots which streaked badly on a paper chromatogram, and traces of glucose were found to be present in the mild acid hydrolysate from which the degraded polymer had been recovered. Further hydrolysis of this hydrolysate at 100° reduced the quantity of material of lower mobility and chromatography showed mannose (28) in addition to the spots already given by the mild acid hydrolysis.

Methylation and periodate oxidation revealed that the degraded polymer consists of chains of about fifteen 1,3-linked mannopyranose units (25) with

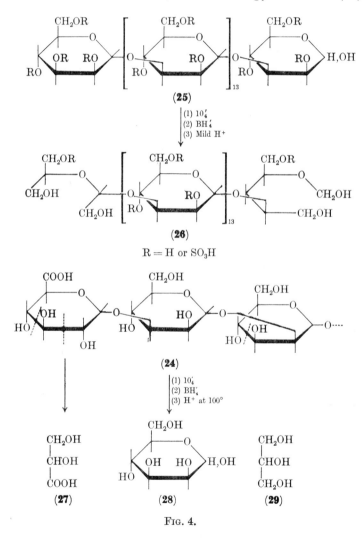

FIG. 4.

occasional residues carrying sulphate groups. This appears to constitute the backbone of the native glucuronomannan to which are attached side chains of the aldotriouronic acid, and of the third oligouronic acid, by as yet unknown linkages. It is considered that the aldobiouronic acid was probably derived by further hydrolysis of the other two oligosaccharides since both these gave rise on further hydrolysis to additional quantities of this disaccharide. Although the initial polysaccharide had a positive rotation indicating the presence of some α-linkages, no evidence for the type of glycosidic linkage in the different fragments was advanced.

During the Smith degradation, the 1,2-linked mannose and the glucuronic acid units in the side chains (24) are oxidized by the periodate and on acid hydrolysis after reduction these are cleaved together with the unoxidized 1,3-linked mannose, and glucose units (from the longer side chains), and are respectively responsible for the glycerol (29), glyceric acid (27), mannose (28) and traces of glucose found in the acid hydrolysate (see Fig. 4).

No glucose could be detected in a total acid hydrolysate of the polysaccharide. It must, therefore, be present in very small amount, and it is unlikely that it has much significance in the overall structure of the molecule.

Any sulphate groups labile to alkali would have been lost during the extraction of this polysaccharide. However, treatment of the residual organism with sodium methoxide before extraction of the glucuronosylmannan should remove any alkali labile sulphate and result in methylation of either the carbon atom, which formerly carried the sulphate residue, or an adjacent carbon atom (see p. 46). Application of this procedure failed to yield a polysaccharide containing any methylated sugars and it seems unlikely, therefore, that any sulphate was lost during the extraction. In the absence of 1,3-linked glucuronic acid, it follows that the sulphate must be linked to mannose since any linked to glucuronic acid would be alkali labile.

This is the first member of this group to be chemically investigated and the first example of this type of polysaccharide, and it is possible that it may prove to be the characteristic cell wall material of this class of organism.

REFERENCES

Barry, V. C., and Dillon, T. (1945). *Proc. R. Irish Acad.* 50B, 349.
Brading, J., Georg-Plant, M. M. T., and Hardy, J. (1954). *J. chem. Soc.* p. 319.
Cessi, C., and Piliego, F. (1960). *Biochem. J.* 77, 508.
Ford, C. W., and Percival, Elizabeth (1965). *J. chem. Soc.* p. 7042.
Haq, Q. N., and Percival, Elizabeth (1966a). *In* "Some Contemporary Studies in Marine Science" (H. Barnes, ed.), p. 355, Allen and Unwin Ltd., London.
Haq, Q. N., and Percival, Elizabeth (1966). *Proc. 5th int. Seaweed Symp.* (1965), Halifax, Nova Scotia (E. G. Young and J. L. McLachlan, eds), p. 261, Pergamon Press, Oxford.

Haug, A. (1964). Report No. 30, Norwegian Institute of Seaweed Research, Trondheim.
Haug, A., and Larsen, B. (1963). *Acta chem. scand.* **17**, 1653.
Larsen, B., Haug, A., and Painter, T. J. (1966). *Acta chem. scand.* **20**, 219.
McKinnell, J. P., and Percival, Elizabeth (1962a). *J. chem. Soc.* p. 2082.
McKinnell, J. P., and Percival, Elizabeth (1962b). *J. chem. Soc.* p. 3141.
O'Donnell, J. J., and Percival, Elizabeth (1959). *J. chem. Soc.* p. 2168.
Percival, Elizabeth (1967). *Chemy. Ind.* 511.
Percival, Elizabeth, and Wold, J. K. (1963). *J. chem. Soc.* p. 5459.
Rees. D. A. (1963). *Biochem. J.* **88**, 343.

CHAPTER 9

Some Comparisons of Algal
With Other Polysaccharides

I. GENERAL DISTRIBUTION OF PLANT POLYSACCHARIDES

With increasing knowledge of the constituents of plants, there has been a growing interest in the distribution of different compounds among plant species, and the possible value of this information in helping to decide outstanding questions on the classification and evolutionary background of the different plant groups. The position regarding many classes of compounds has recently been reviewed (Swain, 1966). The situation with polysaccharides seems to be that there are a number of general types which are common to large groups, and that features limited to smaller groups are to be found only in fine details of structure (Percival, 1966). On the other hand, in some cases, there are wide seasonal and environmental differences to be found in polysaccharides from the same species. Examples from the marine algae are the variations in the 3,6-anhydro-D-galactose content of carrageenan from *Chondrus crispus* (see p. 138) and in the 3,6-anhydro-L-galactose content of porphyran from *Porphyra umbilicalis* (see p. 134), and the changes in the proportion of D-mannuronic and L-guluronic acid residues in the alginate from *Ascophyllum nodosum* and *Laminaria digitata* (see p. 110).

While some polysaccharides are extremely widely distributed, being present in representatives from nearly every class of photosynthetic organism (for example, cellulose and to a lesser extent starch of the amylopectin type), others may be limited to one class (alginic acid in the Phaeophyceae) and others, as far as is known, to relatively few genera (mannans in some members of the Rhodophyceae and Chlorophyceae).

Polysaccharides containing sugar residues with half ester sulphate groups seem to be universal in marine algae, but absent in land plants. The plant sulphatides, widely distributed in green tissues, are sulphonic acids, not ester sulphates (Rees, 1965). On the other hand, O-acetyl groups and methyl esters of uronic acids are found in the polysaccharides of many land plants, but not in those from seaweeds, while O-methylated sugar residues are to be found in the polysaccharides from both types of plant.

II. CONSTITUENT SUGARS IN PHOTOSYNTHETIC PLANTS

Although many sugars are found combined in polysaccharides in all classes of plants, it appears from the material so far examined that some are confined to certain algae or to certain land plants. An indication of their distribution is given in Table 1.

Table 1   Sugars present in some plant polysaccharides.

	Phaeo-phyceae	Rhodo-phyceae	Chloro-phyceae	Flowering plants
D-Glucose	**	**	**	**
D-Galactose	*	**	**	**
D-Mannose	*	**	**	**
L-Galactose		**		*
D-Fructose			**	**
D-Xylose	*	**	**	**
L-Arabinose			**	**
D-Glucuronic acid	*	*	**	**
D-Galacturonic acid		*		**
D-Mannuronic acid	**			
L-Guluronic acid	**			
L-Fucose	**			*
L-Rhamnose			**	*
Mannitol	*			
Sulphate ester	**	**	**	

** Sugars forming a major part of a polysaccharide.
* Sugars found in a polysaccharide in small quantity.

Only the marine species of the Phaeophyceae, Rhodophyceae and Chlorophyceae are considered here, as too few representatives of other algal classes have been investigated to give significant conclusions.

It is considered that the primary photosynthetic process is the same in all plants that have been studied (Brody and Brody, 1962) and that phosphoglyceric acid is a universal intermediate in carbohydrate synthesis. Presumably, enzyme systems have evolved to build up the different sugars and to form from them the wide range of polysaccharides, each of which may provide a certain advantage in its particular environment. There may be interconversion of different sugars, as for example L-galactose from D-mannose (see p. 21), and enzymic oxidation of a sugar nucleotide to the corresponding uronic acid compound (see p. 17) has been demonstrated.

III. POLYSACCHARIDES CONSTITUTING CELL WALLS

A simplified picture of cell walls is that they consist of a polysaccharide microfibrillar framework associated with encrusting or cementing material of a less highly organized nature.

Unfortunately, it is not possible by methods presently available to distinguish clearly between the materials making up the microfibrils and those which hold them together, nor indeed between cell wall substances and materials found between the cells (see p. 14). Although physical methods may give a good idea of the location and form of some of these substances, the problem remains of separating them to carry out the chemical investigations necessary to determine details of structure.

In considering the similarities and differences of the polysaccharides of marine algae and those of land plants, an attempt will be made, where possible, to compare substances having similar functions in the different types of plants, but in view of our very inadequate knowledge in this field, some broad chemical similarities will also be used in making the comparisons.

A. THE MICROFIBRILLAR FRAMEWORK

Cellulose (see p. 83) is by far the most common polysaccharide found as the basis of the cell wall structure. It appears to be universal in land plants and is very common in algae, although in some species the amount is small and it may be less highly polymerized than in land plants. However, there are some algal species which appear to be devoid of cellulose and have another polysaccharide in its place. For example, the structural material has been shown to be a straight chain mannan with β-1,4-links in the red seaweed, *Porphyra umbilicalis* (perhaps outer layer only, see p. 94), and in the green seaweeds of the genera *Codium*, *Derbesia*, *Acetabularia* and *Halicoryne*.

Rhodymenia palmata contains a straight chain xylan and xylans are the main structural materials in the green seaweeds *Caulerpa*, *Bryopsis*, *Halimeda* and *Chlorodesmis* (see p. 91).

B. ENCRUSTING AND INTERCELLULAR SUBSTANCES

Although there are considerable differences in the ease with which these substances can be extracted from the plant tissues, they are all obtained for chemical examination as solutions in water or aqueous acids or alkalis. Although extraction with water can be expected to give products which have suffered little chemical change, the more drastic treatments necessary to obtain many of the others probably results in the material examined chemically being somewhat different from that present in the plant tissues. For example, the presence of some protein fragments in many extracts suggests a more extensive association between the polysaccharides and proteins in the living organisms.

In the higher plants the materials considered are classified as hemicelluloses, pectic substances, gums and mucilages, while the algal extracts are sometimes grouped together as mucilages. A very common, although not universal, feature of these polysaccharides is the presence of acidic groups in the macromolecules. In land plants, the acid groups are present as D-glucuronic and its 4-O-methyl ether or D-galacturonic acid residues, while in the algae both uronic acid residues and half ester sulphates provide acidity. Another very common feature is the presence in the polysaccharide of more than one type of sugar residue.

With the detailed examination of extracts from more and more plant sources, it is becoming apparent that it is unusual for a particular grouping of structural units to be found in only one class of polysaccharide; structural similarities between many hemicelluloses, pectic substances and exudate gums are particularly striking (Aspinall, 1964). Although as far as is yet known some of the extracts from marine algae appear to have structures restricted to a single algal class, the synthesis by bacteria (Linker and Jones, 1966; Gorin and Spencer, 1966) of a polysaccharide similar in properties and structure to alginic acid from the Phaeophyceae, suggests that investigations of more obscure organisms may reveal other similarities.

1. *Hemicelluloses*

The expression hemicellulose is generally restricted to land plants and used to cover the polysaccharides found in close association with cellulose, especially in lignified tissue. The most common method of preparation is by alkali extraction of the "holocellulose" obtained after removal of lignin from the wood. Although all the hemicelluloses contain more than one type of sugar residue, a broad distinction can be made into those containing a high proportion of xylose and those with a high proportion of mannose. The relative amounts of the types of hemicellulose varies considerably with the species of tree from which the wood is obtained (Timell, 1957). The hemicelluloses are all rather different in structure from polysaccharides present in seaweeds.

Xylans are of widespread occurrence in woody tissues of higher plants and are also found in cereals and grasses (Aspinall, 1959). They can be represented by the general structure.

$$\rightarrow 4\text{-}D\text{-Xylp}(\beta 1 \rightarrow 4)\text{-}D\text{-Xylp}(\beta 1 \ldots \rightarrow 4)\text{-}D\text{-Xylp}(\beta 1 \rightarrow$$

$$
\begin{array}{ccc}
3 & 2 & 3 \\
| & | & | \\
1 & 1\alpha & 1 \\
\text{L-Araf} & \text{4Me-}D\text{-GlcUA} & \text{R-L-Araf}
\end{array}
$$

Glc = glucose, GlcUA = glucuronic acid, Xyl = xylose; p and f show pyranose and furanose forms; Me indicates an O-methyl group; R = another chain.

The products from different sources differ in the number of side chains and the proportions of 4-O-methylglucuronic acid and arabinose. Alkali soluble algal xylans are found in association with the "microfibrillar" xylans of green seaweeds and differ from the hemicellulose xylans in that they are essentially β-1,3-linked linear homopolysaccharides. The water soluble xylan from *Rhodymenia palmata* is apparently also a homopolysaccharide, but comprises 80% of 1,4-linked and 20% of 1,3-linked units (see p. 88).

Although combined xylose has been found in other polysaccharides from the red, brown and green seaweeds, in most cases, apart from those mentioned, it comprises only a minor part of the molecule.

The hemicellulose glucomannans, which again are widely distributed among the land plants (Timell, 1965), have a main chain comprising 1,4-β-D-mannose and 1,4-β-D-glucose residues, with, in some cases, galactose units as branches. The ratio mannose:glucose varies but is generally in the range 3:1 to 2:1. Some of the mannan from *Codium* and other seaweeds can be extracted with alkali and is indistinguishable in structure from residual insoluble mannan. This polysaccharide contains, as well as the mannose, about 5% of 1,4-linked glucose residues.

2. *Water-soluble Polysaccharides with Minor Proportions of Uronic Acid*

Arabinogalactans are water-soluble polysaccharides found in considerable amounts in softwoods, particularly larches (Aspinall, 1959; Bouveng, 1961), and in small amounts in some hardwoods (Timell *et al.*, 1958). They are highly branched polysaccharides with a 1,3-linked D-galactopyranose backbone to which are attached a variety of side chains comprising L-arabinofuranose and D-galactopyranose units. A small proportion of D-glucuronic acid units is sometimes present. Polysaccharides of very similar structure are found in exudate gums and it is by no means certain whether the arabinogalactans found in wood, although often referred to as hemicelluloses, are part of the cell wall complex or whether they should be classed among the intercellular compounds.

The arabinogalactans found in the green seaweeds of the genera *Cladophora*, *Chaetomorpha*, *Caulerpa* and *Codium* (see p. 165), although highly branched polysaccharides containing 1,3-linkages, differ from those of land plants in being sulphated. Although these mucilages can be partially extracted from the algae by hot water, further extractions with acid and alkali yield polysaccharides of essentially similar composition.

The galactans which are present as major components of most species of the Rhodophyceae are extracted from the plants by hot or cold water (see p. 127) and differ from those of the land plants in containing alternating 1,3- and 1,4-linkages with either a 6-sulphate or a 3,6-anhydro bridge in the 4-linked residues. Further sulphate ester groups are frequently present, and small proportions of uronic acid are found in galactans from some species.

Fucoidan (see p. 157), found universally in the Phaeophyceae, with its main constituent sugar L-fucose, its predominantly 1,2-linkage and its high degree of sulphation, is unlike any polymer found in land plants. It is typically extractable with water. Some similar polysaccharides which also contain xylose and glucuronic acid and some protein are also found in the Phaeophyceae; while some are readily extracted by alkali, others appear to be closely associated with cell walls and remain in the residue after alkaline extraction.

3. *Polysaccharides Containing a Major Proportion of Uronic Acid*

Pectins, widely distributed in land plants, are characterized by a main chain of 1,4-linked galacturonic acid units. At first sight this suggests a marked similarity to alginic acid present in the Phaeophyceae (see p. 99), but closer examination reveals considerable differences. One which affects the solubility, and hence method of extraction (water or dilute acid for pectin, alkali for alginic acid), is that in most pectins many of the uronic acid groups are present as the methyl ester, while no esterification has been detected in native alginic acid. After de-esterification of pectin to give pectic acid, its properties are in many ways very similar to those of alginic acid, but considerable chemical differences remain. Studies on alginic acid have shown the presence of only 1,4-linked D-mannuronic acid and L-guluronic acid. There has been no evidence of branching or of the presence of other sugars in the chain. Xylose and glucuronic acid have been found in samples of alginic acid but it is probable that they are present in a separate polysaccharide.

On the other hand, pectins from different plants have been shown to contain rhamnose, galactose, xylose, arabinose and fucose in varying amounts. These sugars form part of the main chain in some pectins and branch chains in others (Aspinall, 1964). In addition, the pectins always appear to be closely associated with neutral arabans (Hirst and Jones, 1948) and galactans (Hirst *et al.*, 1947). In some cases, the neutral polysaccharides may form part of a larger polymer which is broken down during extraction, but it has been shown that in mustard seed cotyledons an araban occurs as an independent polysaccharide in addition to the pectin (Rees and Richardson, 1966).

Glucuronic and galacturonic acids are found in different amounts in the exudate gums all of which contain in addition one or more neutral sugars. D-Galactose is commonly present in considerable amounts; L-arabinose is frequently associated with D-glucuronic acid and L-rhamnose with D-galacturonic acid, but different gums provide many combinations of sugars. The gums have highly branched molecules, and among different types, glycosidic linkages to all positions in the sugars have been found.

Polysaccharides of a somewhat similar composition have been isolated from the green seaweeds of the genera *Ulva, Enteromorpha* and *Acrosiphonia* (see p. 178) and also from the only two brown seaweeds (*Ascophyllum nodosum* and *Laminaria hyperborea*) which have been examined for such compounds (see p. 176). The green seaweed extracts contain L-rhamnose, D-xylose and D-glucuronic acid as major components while those of the brown seaweeds contain mainly L-fucose, D-xylose and D-glucuronic acid, in all cases linked to give highly branched molecules. The proportions of the sugars are unlike those so far found in gums, and in addition the seaweed extracts contain half ester sulphate groups.

IV. FOOD RESERVE POLYSACCHARIDES

Until more is known about the metabolism of the seaweeds it will not be possible to say which of the polysaccharides present in them are food reserve materials, but it is not unreasonable to consider that the starches of the red and green seaweeds (see p. 74) and laminaran of the brown seaweeds (see p. 53) have this function.

The relationship of the starches to those of land plants has already been discussed and their essential similarities pointed out.

β-1,3-Linked glucans are widely distributed in plant tissues (Clarke and Stone, 1963), but only in laminaran has the presence of a mannitol end group been detected.

In the Compositeae, the main food reserve is inulin, a straight chain of 2,1-linked β-fructofuranose residues terminated by a sucrose unit at the so-called reducing end, while levans, found in the leaves and stems of many monocotyledons, is built up from 6,2-linked β-fructofuranose residues. Fructans have been isolated from some green seaweeds (see p. 82) and are thought to be inulins, but no proof of their structure has yet been given.

It is probable that the galactomannans found in many leguminous seeds (Smith and Montgomery, 1959) and the glucomannans of tubers of species of Amorphophallus (Smith and Srivastava, 1956) and some other plants are food reserve materials. They do not appear to have a close parallel in the polysaccharides of the algae.

V. SOME COMPARISONS WITH ANIMAL POLYSACCHARIDES

In view of the wide biological difference between algae and multicellular animals, it is remarkable that polysaccharides with half ester sulphate groups are found in both these groups of organisms while being absent from flowering plants. As in the algae, sulphated polysaccharides are found in spaces between animal cells, and may be associated with cell walls. It has been suggested that the effect of λ-carrageenan in stimulating the growth of connective tissue is a consequence of similarities in structure (see p. 154), and the similarity of behaviour of heparin from animal tissues and fucoidan from algae in inhibiting the clotting of blood is also of interest (see p. 164).

On the other hand, the sulphated polysaccharides found in animal tissues appear always to contain some amino sugar residues, and the sulphate ester groups are frequently attached to these residues (Rees, 1965), whereas in polysaccharides from algae no proof of the presence of amino sugars has been sustained.

The similarity of the salt content of the serum of higher animals to that of primitive oceans has been advanced as part of the chemical evidence for the origin of animal life in the sea, and the presence of sulphated polysaccharides in animal tissues may point in the same direction.

REFERENCES

Apsinall, G. O. (1959). *In* "Advances in Carbohydrate Chemistry", Vol. 14, p. 429, Academic Press, New York and London.

Aspinall, G. O. (1964). *In* "Progress and Prospects in the Chemistry of Cell Wall Polysaccharides" (*in* Chemie et Biochemie de la Lignine, de la Cellulose et des Hemicelluloses), Symposium International, p. 421, Grenoble, 1964.

Brody, M., and Brody, S. S. (1962). *In* "Physiology and Biochemistry of Algae" (R. A. Lewin, ed.), p. 3, Academic Press, New York and London.

Bouveng, H. O. (1961). *Svensk kem. Tidskr.* **73**, 113 and ref. cited therein.

Clarke, A. E., and Stone, B. A. (1963). *Rev. pure appl. Chem.* **13**, 134.

Gorin, P. A. J., and Spencer, J. F. T. (1966). *Can. J. Chem.* **44**, 993.

Hirst, E. L., and Jones, J. K. N. (1948). *J. chem. Soc.* p. 2311.

Hirst, E. L., Jones, J. K. N., and Walder, W. O. (1947). *J. chem. Soc.* p. 1225.

Linker, A., and Jones, R. S. (1966). *J. biol. Chem.* **241**, 3845.

Percival, Elizabeth (1966). *In* "Comparative Phytochemistry" (T. Swain, ed.), p. 139, Academic Press, London and New York.

Rees, D. A. (1965). *Rep. Progr. Chem.* **62**, 469.

Rees, D. A., and Richardson, N. G. (1966). *Biochemistry* **5**, 3099.

Smith, F., and Srivastava, H. C. (1956). *J. Am. chem. Soc.* **78**, 1404.

Smith, F., and Montgomery, R. (1959). "The Chemistry of Plant Gums and Mucilages", Reinhold, New York.

Swain, T. (1966). "Comparative Phytochemistry", Academic Press, London and New York.

Timell, T. E. (1957). *TAPPI* **40**, 568.

Timell, T. E. (1965). "Cellular Ultrastructure of Woody Plants", Syracuse University Press, Syracuse, New York.

Timell, T. E., Glaudemans, C. P. J., and Gillham, J. K. (1958). *Pulp Pap. Can.* **59**, No. 10, 242.

Author Index

Kristensen, K., 157, *174*
Krotkov, G., 16, 19, *24*
Kylin, H., 73, 85, *97*, 157, *174*

L

Larsen, B., 26, *51*, 100, 103, 105, 108, 109, 111, 115, 122, *124*, *125*, 138, 150, *156*, 158, *174*, 176, *189*
Lawley, H. G., 56, *72*
Lehoczky-Mona, J., 154, *155*
Leonard, V. G., 69, *72*
Lester, R., 66, 67, *72*
Levring, R., 2, 8, *25*
Lewes, B. A., 35, 36, 39, 41, *51*
Lid, I., 112, *124*
Lin, T. Y., 16, 17, 18, 19, 20, *25*
Lindberg, B., 4, 5, *25*, 33, 41, *51*, 90, *96*
Linker, A., 113, *125*, 193, *197*
Lipatov, S. M., 146, *155*
Little, A. H., 85, *97*
Lloyd, A. G., 48, *51*
Lloyd, K. O., 88, *97*, 157, *174*
Love, J., 11, 14, *25*, 27, 28, *51*, 80, 95, *97*, 165, 171, *174*
Lucas, H. J., 102, 120, *125*
Luchsinger, W. W., 67, 68, *72*
Lund, S., 129, *155*
Lunde, G., 107, *125*, 157, 159, 163, *174*

M

Maass, H., 101, 123, *125*
McCandless, E. L., 154, *155*
McCormick, J. E., 75, 89, *96*, 145, *154*
MacCraith, D., *155*
McCully, M. E., 12, *25*, 157, *174*
McDowell, R. H., 27, *52*, 104, 117, 120, 123, *125*
McInnes, 2, *24*, *25*
McKenna, J., 145, *155*
Mackie, I. M., 11, *25*, 28, 49, *52*, 53, *72*, 79, 91, *97*, 170, 171, *174*
Mackie, W., 28, *51*, 80, *97*, 165, 166, 168, *174*

McKinnell, J. P., 11, *25*, 80, 87, *97*, 165, *174*, 179, 180, *189*
McLachlan, J. L., 2, 5, 6, 15, *24*, *25*
McNeal, G. M., Jr., 164, *175*
McNeely, W. H., 99, 101, 120, 121, *125*, 163, 164, *174*
Macpherson, M. G., 6, *25*
Majak, W., 5, 6, 15, *24*, *25*
Makita, M., 31, *52*
Mandels, M., 67, 68, *72*
Manners, D. J., 20, 22, *24*, *25*, 43, 44, 45, *52*, 57, 58, 60, 61, 63, 65, 66, 67, *71*, *72*, 73, 76, 77, 89, 90, *97*, 164, *174*
Marechal, L. R., 19, *24*, *25*
Marshall, S. M., 150, *155*
Martin, W. G., 64, *72*
Massoni, R., 103, *125*
Mead, T. H., 130, *156*
Meeuse, B. J. D., 4, 12, *25*, 53, *72*, 77, 78, 79, 82, 83, *97*
Meier, H., 28, *52*, 170, *174*
Mérac, M. L. Rubat, du, 82, *97*
Meredith, W. O. S., 67, *71*
Meyer, K., 113, *125*
Mitchell, J. P., 22, *25*, 89, 90, *97*
Mitchell, P. W. D., 75, 89, *96*
Miwa, T., 11, 14, *24*, *25*, 45, *51*, 91, 92, 95, 96, *97*, *98*
Molland, J., 112, *124*
Mongar, I. L., 119, *125*
Montgomery, R., 196, *197*
Morgan, K., 140, *155*, 164, *175*
Mori, T., 144, *155*
Morozov, A. A., 146, *155*
Moscatelli, E. A., 67, *72*
Moss, B. L., 7, 10, *25*
Myers, A., 12, *24*, 84, 86, 87, *97*
Myklestad, S., 106, 115, *125*

N

Nace, G. W., *51*, 130, *155*
Naylor, G. L., 85, 86, *97*
Neal, J. L., 140, 146, *156*
Nelson, W. L., 101, *125*, 163, *175*
Neufeld, E. F., 16, *25*
Nevell, T. P., 30, *52*
Newton, L., 150, *155*
NiOlain, R. M., *155*

Subject Index

A

Acanthopeltis, agar, 128, 130
Acanthopeltis japonica
  agarose, proportion in, 130
    composition of, 132
Acetabularia calyculus
  mannan, 94
Acetabularia mediterranea
  fructan, 82
Acetolysis of fucoidan, 162
Acrosiphonia arcta
  glucan, 179
  sulphated polysaccharide, 179
    composition, 179
    hydrolysis, partial, 180
    methylation, 180
    periodate oxidation, 179
Acrosiphonia centralis
  see *A. arcta*
Agar, 127, *129–134*
  applications, 126, *152*
  Danish, 143
  definition, 128, 152
  distribution, 128
  enzymic hydrolysis, 131, 146
  fractionation, 129
  isolation, 129
  molecular weight, 146
  properties, 149
  structure, *129–134*
Agarase 131, 136
  porphyran, action on, 136
  purification, 148
Agarobiose, 131, 134
  diethyl dithioacetal, 130, 131
  dimethylacetal, 130, 131, 134
  glycoside, 130, 131

6-O-methyl, dimethyl acetal, 131
  synthesis, 131
*neo*Agarobiose, 45, 131
Agaroids, 128
Agaropectin, 129, 147
  enzymic hydrolysis, 133
  oligosaccharides from, 134
  structure, *133–134*
Agarose, 129, 147
  applications, 152
  properties, 150
  structure, *130–133*
Agarotetraose dimethyl acetal, 134
Ahnfeltia
  agar, 128
  cellulose, 87
  galactans, 128
Ahnfeltia plicata
  agaropectin, 133
  L-arabinose in, 133
Alaria esculenta
  alginate composition, 110
  laminaran content, 53
Alginase, 112
Alginic acid, *93–123*
  alkali on, 122
  applications, 122
  bacterial, 193
  base exchange reactions, 118
  biosynthesis, 16–22
  composition of, in different algae, 110
  constitution and structure, *101–106*
  degradation, 121
  derivatives, 120
  determination of, 108
  dissociation constant, 114
  distribution, 3, 99
  enzymic hydrolysis, 45, *112–114*

enzymic hydrolysis, 136, 148
structure, 134–137
Prasiola japonica
cell wall, 14
mannan, 96
Protein in
Acrosiphonia, 180
ascophyllan, 176, 177
Chaetomorpha, 169
Cladophora, 169
Enteromorpha, 180
fucoidan, 159
glucuronoxylofucan, 176, 177
Ulva, 180
Protein
metabolism of, 8
removal, 26
Pseudodichotomosiphon
xylan, 91
Pterocladia tenuis
agarose proportion, 130
composition, 132
Pterocladia spp.
agar, 128
Pterocladia pyramidale
cellulose, 87
Ptilota plumosa
cellulose, 87
pyruvic acid in
agaropectin, 133

R

Red seaweeds
see Rhodophyceae
Respiration, 22
L-Rhamnose in polysaccharides from
Acrosiphonia, 179, 180
Chaetomorpha, 168
Cladophora, 165
Enteromorpha, 179
plants, 191
Ulva, 179
Rhizoclonium
starch, 79
Rhodochorton floridulum, cellulose, 87
Rhodoglossum affine, 87
Rhodophyceae, 3
see also under individual species

carbohydrates, 5, 191
cellulose, 87
classification, 128
galactans, 127–124
mannans, 93
polyuronides, 176
starch, 73–78
xylan, 88
Rhodymenia palmata
cellulose, 87
enzymes from, 68, 164
transglucosylation in, 20
xylan, 3, 88–91

S

Sargassum linifolium
alginate composition, 110
Scytosiphon lomentaria, alginate composition, 110
Smith degradation of
λ-carrageenan, 141
laminaran, 61
Phaeodactylum polysaccharide, 186
Ulva polysaccharide, 179
see also periodate oxidation
Solubility
alginates, 114
Spermatochnus paradoxus, alginate composition, 110
Sphacelaria bipinnata, alginate composition, 110
Spongomorpha arcta
see *Acrosiphonia arcta*
Staining reagents, 12
Starches, 3, 4, *73–82*
see also Chlorophyceae starch and floridean starch
amylopectin, 75, 78, 79, 81
amylose, 79, 80
comparison, 76, 80, 81
Cyanophyceae, 82
distribution, 3, 4, 76, 77, 79
granules, 78, 79, 82
identification, 12
periodate oxidation, 77, 80, 81
Structural studies
see also periodate oxidation and specific polysaccharides